The Corresponding-States Principle and its Practice

Thermodynamic, Transport and Surface Properties of Fluids

The Corresponding-States Principle and its Practice
Thermodynamic, Transport and Surface Properties of Fluids

By

Hong Wei Xiang

Institute of Chemistry
Chinese Academy of Sciences
Beijing
China

2005
ELSEVIER

Amsterdam – Boston – Heidelberg – London – New York – Oxford
Paris – San Diego – San Francisco – Singapore – Sydney – Tokyo

014148547
phys

ELSEVIER B.V.   ELSEVIER Inc.   ELSEVIER Ltd.   ELSEVIER Ltd.
Radarweg 29   525 B Street, Suite 1900   The Boulevard, Langford Lane   84 Theobalds Road
P.O. Box 211, 1000 AE Amsterdam   San Diego, CA 92101-4495   Kidlington, Oxford OX5 1GB   London WC1X 8RR
The Netherlands   USA   UK   UK

First edition 2005

Library of Congress Cataloging in Publication Data
A catalog record is available from the Library of Congress.

British Library Cataloguing in Publication Data
A catalogue record is available from the British Library.

ISBN: 0-444-52062-7

♾ The paper used in this publication meets the requirements of ANSI/NISO Z39.48-1992 (Permanence of Paper).
Printed in The Netherlands.

NM   CC²
       11/27/05

On the Occasion of the 125th Anniversary since
Johannes Diderik van der Waals discovered the
Corresponding-States Principle
in Amsterdam in 1880

To

All the Hearts that Directly and Indirectly Contributed
to this Book

Presentation Speech at the Ceremony of the Nobel Prize to Johannes Diderik van der Waals

As far back as in his inaugural dissertation "The Relationship between the Liquid and the Gaseous State", Van der Waals indicated the problem to which he was to devote his life's work and which still claims his attention today. In the dissertation to which I have referred, he sought to account for the departure from the simple gas laws which occur at fairly high pressures.......However, this was not the most important result of Van der Waals' studies. His calculations led him to consider that once we are acquainted with the behavior of a single type of gas and the corresponding liquid, e.g. that of carbon dioxide, at all temperatures and pressures, we are able by simple proportioning to calculate for any gas or liquid its state at any temperature and pressure, provided that we know it at only one, *i.e.*, the critical temperature.

On the basis of this law of what are known as "corresponding states" for various gases and liquids Van der Waals was able to provide a complete description of the physical state of gases and, more important, of liquids under varying external conditions. He showed how certain regularities can be explained which had earlier been found by empirical means, and he devised a number of new, previously unknown laws for the behavior of liquids......Van der Waals' theory has also been brilliantly successful through its predictions which made it possible to calculate the conditions for converting gases to liquids. Two years ago Van der Waals' most prominent pupil, Kamerlingh Onnes, in this way succeeded in compelling helium — the last previously uncondensed gas — to assume the liquid state.

Yet Van der Waals' studies have been of the greatest importance not only for pure research. Modern refrigeration engineering, which is nowadays such a potent factor in our economy and industry, bases its vital methods mainly on Van der Waals' theoretical studies.

—— by the Rector General of National Antiquities, Professor O. Montelius, President of the Royal Swedish Academy of Sciences, on December 10, 1910 from *Nobel Lectures. Physics 1901-1921*, Elsevier, 1967.

Comment on the Importance of the Corresponding-States Principle

This law had a particular attraction for me because I thought to find the basis for it in the stationary mechanical similarity of substances and from this point of view the study of deviations in substances of simple chemical structure with low critical temperatures seemed particularly important.

—— H. Kamerlingh Onnes, Nobel lecture on December 11, 1913

The principle of corresponding states may safely be regarded as the most useful byproduct of Van der Waals' equation of state. Whereas this equation of state is nowadays recognized to be of little or no value, the principle of corresponding states correctly applied is extremely useful and remarkably accurate.

—— E. A. Guggenheim at Imperial College at University of London,
J. Chem. Phys. 13: 253, 1945

The most generally useful method of prediction of the volumetric properties of fluids is the hypothesis of corresponding states, which also came originally from Van der Waals. Engineers have used this method extensively to obtain estimated properties for design purposes, and many authors have presented charts of both volumetric and related thermodynamic properties on the basis of corresponding states.

—— K. S. Pitzer at University of California at Berkeley,
J. Am. Chem. Soc. 77: 3427, 1955

The most powerful tool available for quantitative prediction of the physical properties of pure fluids and mixture is the corresponding states principle.

—— T. W. Leland and P. S. Chappelear at Rice University,
Ind. Eng. Chem. 60(7): 15, 1968

I was confronted with the precise application of thermodynamic methods, and I could understand their usefulness......I mean the solutions theory, the theory of corresponding states and of isotopic effects in the condensed phase.

—— I. Prigogine, *Les Prix Nobel, Chemistry 1971-1980,* 1977.

There is one theory, which is particularly helpful. The successful application of the law of corresponding states has encouraged many correlations of properties, which depend primarily on intermolecular forces. Many of these have proved invaluable to the practicing engineer.

—— R. C. Reid and T. K. Sherwood at MIT, J. M. Prausnitz at University of California at Berkeley, B. E. Poling at University of Toledo, and J. P. O'Connell at University of Virginia, *Properties of Gases and Liquids*, 3th, 4th, 5th eds., McGraw-Hill, New York, 1977, 1987, 2001.

The practical usefulness of the corresponding-states theory can hardly be overemphasized. It forms the basis of numerous correlations of physical properties.

—— T. M. Reed and K. E. Gubbins at North Carolina State University, *Applied Statistical Mechanics: Thermodynamic and Transport Properties of Fluids,* Butterworth-Heinemann, Stoneham, 1991.

The most important work I did at Berkeley was on Pitzer's extension of the Theory of Corresponding States.

—— R. F. Curl, Jr., *Les Prix Nobel, Chemistry* 1996.

The general corresponding-states principle has proved to be much better than has often been thought in the past. The methods based on the principle are theoretically based and predictive. It has a firm basis in statistical mechanics and kinetic theory, and has a great range and accuracy. It should not only be able to represent data to a reasonable degree but, more importantly it does what a correlation cannot do——predict the properties beyond the range of existing data.

—— M. L. Huber and H. J. M. Hanley at NIST, in *Transport Properties of Fluids,* Cambridge University Press, 1996.

The most powerful tool available today (just as 25 years ago) for making highly accurate, yet mathematically simple, predictions of thermophysical properties of fluids and fluid mixtures is the corresponding states principle......Its fundamentals and applications to pure-fluid and mixtures have been reviewed in almost all of the recently published thermodynamic and statistical mechanics books......All modern generalized engineering equations of state are examples of applications of this principle.

—— J. F. Ely and I. M. F. Marrucho at Colorado School of Mines, in *Equations of State for Fluids and Fluid Mixtures*, Elsevier, Amsterdam, 2000.

Comments on the Extended Corresponding-States Theory of Highly Nonspherical Molecules, on which this book focuses

This paper predicts a general scheme for a wide set of substances and demonstrates a close fit with the available predictive methods —— Reviewer's comments from "The new simple extended corresponding-states principle: complex molecular transport properties in dilute gas state," **Fluid Phase Equilibria, 187** (2001) 221-231.

Interesting results on a new extended corresponding-states principle, which seems to be superior to older formulations. This could be a result of considerable practical importance and is certainly a valuable contribution to the International Journal of Thermophysics —— Reviewer's comments from the "Vapor pressures from a corresponding-states principle for a wide range of polar molecular substances," **International Journal of Thermophysics, 22** (2001) 919-932.

Programmed up your equation and everything looks very good —— Reviewer's comments from the "Vapor pressures, critical parameters, boiling points, and triple points of halomethane molecular substances," **Journal of Physical and Chemical Reference Data, 30** (2001) 1161-1197.

This is a very well written and interesting paper which I can recommend for publication in the Chemical Engineering Science ——Reviewer's comments from "The new simple extended corresponding-states principle: vapor pressure and second virial coefficient," **Chemical Engineering Science, 57** (2002) 1439-1449.

These are simple and accurate and need to be applied more widely. Also this general approach should be extended to other thermophysical properties —— **Dr John H. Dymond**, University of Glasgow, UK.

A full understanding of the topic and of the underlying engineering basis for the theory and its industrial application —— **Professor William A. Wakeham**, Vice Chancellor, University of Southampton, UK.

This extended corresponding-states theory includes highly nonspherical polar, associating, and hydrogen-bonding molecules into the corresponding-states framework with about one order of magnitude more accuracy than that of Pitzer et al. —— **Professor Bu-Xuan Wang**, Tsinghua University, China.

It provides a new and somewhat different approach to a corresponding-states model. It is probably about time that a new approach is attempted —— **Late Dr Lloyd A. Weber**, National Institute of Standards and Technology (NIST),

Contents

The Corresponding-States Principle

The Corresponding-States Practice

PREFACE

On the occasion of the completion of this book, for which I would like to write a few words to express my intentions, I hope to help researchers to understand the current progress in this field, technical workers to learn how to derive practical equations from the molecular modeling, and practicing engineers to find it suitable as a useful handbook.

This book elaborates an old and yet still new topic in thermodynamics, the corresponding-states principle, which has continued to be useful for over a century, since the time of Van der Waals. In many years of engineering practice, the corresponding-states principle has helped us understand and calculate the thermodynamic, transport, and surface properties of substances in various states, which are required by our modern lifestyle. Completed by Pitzer et al on the other side of the Pacific just before half a century, the corresponding-states theory for weakly nonspherical molecules or so-called normal fluids is now extended to highly nonspherical molecules that include polar, hydrogen-bonding and associating substances. The origins and applications of the corresponding-states principle are described from a universal point of view with comparisons to experiments when possible.

Thermodynamics is becoming increasingly more understood through the use of statistical mechanics and molecular simulations, which are expected to sufficiently develop for practical applications. The theme of this book is to use the universal theory, the corresponding-states principle, to explain the present theories and knowledge; although it is impossible to include all the tremendous amount of literature that has been published to date in such a limited book. In the book, I have always felt that we, as scientific researchers, keep in mind that society's needs advance scientific research, which, in turn, serves the society. Therefore, the book seeks to theoretically clarify each question within the limits of our current knowledge for practical applications. It is also strived to help the reader realize the gap between what we have obtained and the final goal we would attain. For the corresponding-states methods to be successfully applied at the present and in the near future, molecular thermodynamics should be characterized by a combination of empirical models for specific properties supported by molecular theory and verified by fundamental experimental data. While the emphasis is on the properties of pure systems, the corresponding-states theory can also be extended to mixtures, which are treated as pure systems.

The work on the extended corresponding-states theory of highly nonspherical molecules has been commented by Chun Li Bai, J. Bridgwater, R. A. Brown, M. Chase, Dai Jing Cheng, D. M. Christopher, R. F. Curl, U. K. Deiters, J. H. Dymond, J. F. Ely, A. Foughour, D. G. Friend, F. Gladden, A. Goodwin, W. M. Haynes, M. L. Huber, E. Kiran, A. Laesecke, E. W. Lemmon, A. Lenhoff, J. M. H. Levelt Sengers, Jing Hai Li, Yun Li, J. W. Magee, G. A. Mansoori, K. N. Marsh,

M. Radosz, M. de Reuck, J. M. Prausnitz, S. I. Sandler, M. Schmid, J. V. Sengers, G. S. Soave, Lian Cheng Tan, J. P. M. Trusler, W. A. Wakeham, Bu Xuan Wang, the late L. A. Weber, J. Willson, Jian Xu, and Ming Shan Zhu to whom I am deeply grateful. As the work has taken into this book, helpful comments and discussions are also contributed by some of these, who read all or part of the manuscript, to whom I am also deeply grateful. The excellent comment from the publisher, especially my editor M. S. Thijssen, is acknowledged. The financial support from the President Fund of the Chinese Academy of Sciences, the Director Fund of Beijing Substance Science Base, Departments of Education and of Personnel, and National Natural Science Foundation of China is also acknowledged.

As poetized by Yuan Ming Tao (363-427), a remarkable work should be shared and its subtleties be discussed. The book does not dare to describe itself as remarkable, yet I hope the readers to discuss and correct if necessary, since it is impossible to avoid negligence and errors due to limited time and my learning.

Hong Wei Xiang
Beijing

Chapter 1

Introduction

1.1 Overview of the Corresponding-States Principle

1.1.1 The Corresponding-States Theory of Monatomic Molecules

As the properties of substances are determined by the behavior of the molecules in these substances, the behavior of the molecules must be understood to know the properties of substances. Many molecular theories have been proposed to describe the properties of substances. One of these, the corresponding-states theory, is especially useful. For more than one hundred years, the corresponding-states principle has been used as the most useful, most reliable and most universal theory.

Proposed by Van der Waals in 1880, the corresponding-states theory for monatomic molecules expresses the generalization that properties that depend on intermolecular forces are related to the critical properties in a universal way. The corresponding-states principle provides the most important basis for the development of correlations and estimation methods for properties. The relation of the intermolecular potentials of monatomic and symmetrical atomic molecules to the corresponding-states properties shows that spherical molecules conform well to the substance similarity principle, upon which the macroscopic laws of the corresponding-states principle and statistical mechanics were established. Van der Waals showed the corresponding-states principle to be theoretically valid for all pure substances whose pressure-volume-temperature properties may be expressed by a two-parameter equation of state. For example, Figure 1.1 illustrates the corresponding-states form of the vapor-liquid coexistence curve of some typical substances for different classes of molecules, where $T_r = T / T_c$ is the reduced temperature, T_c is the critical temperature, $V_r = V / V_c$ is the reduced volume and V_c is the critical volume. However, the two-parameter corresponding-states theory proposed by Van der Waals can not exactly represent the behavior of other classes of molecules that are not spherical.

The successful application of the corresponding-states principle to property correlations has led to similar correlations for properties that depend primarily on intermolecular forces that are invaluable in practical applications. The corresponding-states principle was originally macroscopic; however, its modern form has a molecular basis. As verified by Pitzer in 1939, it is similarly valid if the intermolecular potential function requires only two characteristic parameters.

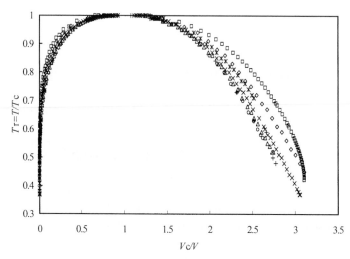

Fig. 1.1 Vapor-liquid coexistence curves of classes of typical substances (○) Argon; (+) Methane; (Δ) Nitrogen; (×) Ethylene; (*) Carbon dioxide; (◊) 1,1,1,2-Tetrafluoroethane; (□) Wwater

1.1.2 The Corresponding-States Theory of Weakly Nonspherical Molecules

The corresponding-states principle can be derived from statistical thermodynamics when severe simplifications are introduced into the partition function. Other useful results can be obtained by introducing less severe simplifications into the statistical thermodynamic equations so as to provide a framework for developing estimation methods. The fundamental equations describing various thermodynamic and transport properties can be derived provided that an expression is available for the potential energy function for the molecular interactions. Such a function may be, at least in part, empirical; however, the fundamental equations for properties are often insensitive to the details in the potential function from which they stem, and two-parameter potential functions usually work remarkably well for various systems.

Spherically symmetric molecules conform to a two-parameter

corresponding-states principle very well; while weakly nonpolar and weakly polar molecules, the so-called normal fluids defined by Pitzer et al., have deviations are often large enough to require another third parameter in correlations. Pitzer et al. proposed the acentric factor to represent the deviation of the vapor pressure-temperature relation from which might be expected for a similar substance consisting of spherically symmetric molecules. The three-parameter corresponding-states correlation of Pitzer et al., when expressed in terms of dimensionless properties, works very well for nonpolar molecules. As a result, good correlations exist for the properties of nonpolar small molecules from the corresponding-states theory of Pitzer et al., which was derived from macroscopic properties but with a molecular basis, correlates the properties of weakly nonspherical molecules with an accuracy that is about one order of magnitude better than that of the Van der Waals corresponding-states theory, however it can not be applied to describe the properties of highly nonspherical molecules.

1.1.3 The Corresponding-States Theory of Highly Nonspherical Molecules

The three-parameter corresponding-states theory for weakly nonspherical molecules cannot describe properties of highly nonspherical polar molecules. The generalized corresponding-states theory described in this book is an important contribution to the completion of the corresponding-states principle, in which the corresponding-states theory of Pitzer et al. for normal fluids is extended to highly nonspherical complex systems of polar, associating and hydrogen-bonding molecules. The extended corresponding-states parameter in this theory has been defined in terms of the deviation of the critical state of a molecule from that of spherical molecules and may physically describe the effects of differences between the structure of nonspherical molecules and that of spherical molecules. The extended corresponding-states parameter is not derived from experimental data for a specific property, so it may be universally used to represent the properties of simple, nonpolar, polar, hydrogen-bonding, and associating molecules. The extended corresponding-states theory provides a new approach to elucidate the physical behavior of nonspherical molecules. The extended corresponding-states theory is generally applicable to all classes of substances of various chemical structures and is a universal theory applicable to all physical properties which provides a reliable basis for the studies of molecular thermodynamics and other related sciences.

1.2 Properties of Substances

1.2.1 Importance of Properties of Substances

Substances are important in all aspects of our lives and for social progress. Substances also play an important role in various industries, production, and operations in our modern lives. Some substances that are of great significance to industrial processes are water, carbon dioxide, air, and liquid and gaseous fuels. In engineering practice, the properties of substances are the basis for analysis and design in many industries, such as energy, power, petrochemical, metallurgy, refrigeration, and heating.

The use of existing substances and of new substances requires reference data for the properties of those substances. As a result, these properties must be understood. Even theoretical physicists need to know the properties of substances to verify their theories by comparison with experimental results. An enormous amount of data has been collected and correlated over the years, but the rapid advance of technology into new fields always seems to maintain a significant gap between the demand and the availability of property data due to our limited understanding of properties of substances. Consequently, proper understanding of the properties of substances is extremely important for the practical applications.

1.2.2 Necessity of the Prediction of Properties

Practical applications frequently require properties of substances that have not been measured or cannot be calculated from existing theories. Reliable methods are needed to predict the properties of these substances. Many handbooks provide data sources, and more and more journals have been devoting to the compilation and critical review of property data. Furthermore, computational databases have become a routine part of computer-aided process design. However, the number of interesting compounds is very large and that of mixtures formed by these compounds is much larger in practical applications. It has been reported that there are more than 50 million different chemical substances, 20 millions of which are often used. However, only a few dozen substances have relatively complete experimental data. In other cases, the required data cannot be obtained from just experimental measurements. Measurement of the properties of all substances for all conditions is absolutely impossible. Moreover, the increasing need for accurate data has ever further outstupped the accumulation of new data, especially for the data of multi-component mixtures. Even if it is possible to obtain the desired properties from new experimental measurements, the measurements are often not practicable because these are expensive and time-consuming. The labor and expense of experimental measurements is almost always reduced by the ability to predict the required properties. The more applicable equations are more able to reflect the natural behavior and more able to help us understanding the properties of substances that are not yet available. Thus, the need is for better tools to better predict the required properties of substances from the limited available data. Such tools require a universal theory with a theoretical basis that is simple and general in form reliably predict physical properties.

1.3 Organization of the Book

This book is partitioned into two parts: the Corresponding-States Principle and the Corresponding-States Practice. Chapter 2 introduces Van der Waals' corresponding-states theory as related to the continuity of the gas and liquid states and Van der Waals' equation of state. The theoretical basis of the corresponding-states principle including the assumptions, derivation, and the expressions for the corresponding states for spherical and nonspherical molecules are described in Chapter 3. The parameters in the corresponding-states principle, i.e. the critical parameters, the acentric factor, and the aspherical factor, and the basic form of the corresponding-states properties are explained in Chapter 4 to complete the presentation of the corresponding-states principle. In the second part, the corresponding-states principle is applied to the thermodynamic properties in Chapter 5. The equations for the virial coefficients, compressibility factor, enthalpy, entropy, fugacity, and heat capacity are derived from their thermodynamic relations using the extended corresponding-states theory for virial, crossover, cubic, hard sphere, Martin-Hou, and liquid state. The corresponding-states thermodynamic properties over the entire fluid state are presented along with calculated deviations. Corresponding-states vapor pressures are described in Chapter 6. The viscosity and thermal conductivity with their corresponding states in the zero-density, vapor phase, critical region, and the entire fluid state are presented in Chapter 7. The corresponding-states surface tension is treated in Chapter 8.

Corresponding-States Principle

Chapter 2

The Corresponding-States Principle from the Continuity of Vapor and Liquid States

2.1 The Continuity of Vapor and Liquid States

While the kinetic theory of molecules gradually became scientifically rigorous, more precise experiments revealed that the behavior of most gases is inconsistent with that of the ideal gas. In 1847, Regnault proved by a number of experiments that, except for hydrogen, no gas exactly complies with Boyle's law and that the expansion coefficient of a real gas increases with the increase of pressure. In 1852, Joule and Thomson discovered the Joule-Thomson effect, in which the internal energy of a real gas changes during its expansion. This experiment verified the existence of the interaction forces between molecules. In 1860, when the Russian chemist Mendeleev investigated the relationship between surface tension and temperature, and it was found that the vapor-liquid interface tension disappears at a specific temperature where the vapor and liquid phases become a single phase. Mendeleev called this temperature the Absolute Boiling Temperature. In 1869, Andrews showed that it was impossible to liquefy carbon dioxide when the temperature is higher than $31.3\,°C$.

In his Lecture on the Continuity of the Gaseous and Liquid States of Matter to the Royal Society, Andrews (1869) described his detailed study of carbon dioxide. He showed that by suitable changes of pressure and temperature it is possible for matter to pass from a gas phase to a liquid phase. He deduced that the ordinary gaseous and liquid states are only widely separated forms of the same condition of matter and can be made to pass from one form to another by a series of gradations so gentle that the passage does not cause any interruption or breach of continuity. In 1871, James Thomson proposed a further viewpoint to interpret Andrews' experiments. Thomson suggested that the sharp breaks at the vapor and liquid ends of isotherms below the critical temperature could be avoided. However, his report provided neither a quantitative calculation nor an explanation from the theory of molecules.

2.2 The Van der Waals Equation of State

Van der Waals, who was studying for his doctoral thesis in Mathematics and Physics at the University of Leiden, employed the kinetic theory of gases, which was being developed at that time, to explain the experiments of Andrews and the review of James Thomson. To describe the vapor-liquid phase transition, Van der Waals suggested that two basic molecular interactions should be considered. First, each molecule has the volume b and, as a result the effective volume for molecular movement decreases to $V - b$, which means that the molecules repel one another on close, but not infinitely close, approach. Second, the attractive effect between the molecules is enough to compensate for the internal pressure, which increases p to $p + a/V^2$. In addition, he derived further that b is equal to four times the volume of the molecules. The critical parameters may be derived from the parameters a and b in the equation of state. Van der Waals wrote his doctoral thesis in Dutch, which quickly received much attention, since Maxwell noticed it and further reviewed his thesis in *Nature*.

As early as in the eighteenth century, Bernoulli suggested that the corrected term b should be introduced into the volume in the equation of state for the ideal gas,

$$p(V - b) = RT , \tag{2.1}$$

in which b is the sum of the volumes of the molecules. In 1863, Hirn suggested the following equation of state:

$$(p + \psi)(V - b) = RT . \tag{2.2}$$

He realized that ψ might be a function of volume and greatly exceeded p in the liquid state, which enlightened Van der Waals later.

When considering the force on any given particle, it is only necessary to take account of the other particles that are within a sphere of a very small radius having the particle as the center. This area is termed the sphere of action, and the forces themselves become insensible at distances greater than the radius of the sphere. If the density is the same throughout, it follows that all those points will be in equilibrium around which we can describe a sphere of action without encroaching on the boundary. Of course this means that the particles will be in equilibrium. These last particles lie on the boundary and form a layer whose thickness is the radius of the sphere of action; and the forces on these particles are directed inwards. One could consider an infinitely thin column in the boundary layer and imagine a part of space below this layer, within the body, containing every molecule that could attract the column. If in this space there were a molecule at rest, it would be necessary to know the law of force to be able to estimate its attraction on the column. But if this molecule is in motion, and can occupy any part of the space indifferently, the above difficulty for the most part disappears, and we can take the attraction exerted by the molecule to be the mean of the attractions, which it would exert in its different possible positions in the space. The same consideration applies to a second molecule, which may be within the space at the same time as the first. In short, the attraction exerted by the matter in the space mentioned is proportional to the quantity of matter, or to the density. The same holds for the molecules within the column, so that the attraction is proportional to the square of the density, or inversely proportional to the square of the volume. Consequently, Van der Waals obtained the equation of state as follows:

$$(p + \frac{a}{V^2})(V - b) = RT. \tag{2.3}$$

This equation of state can be derived from the potential energy function. Van der Waals further investigated the relation of b with the molecular volume.

The reason Van der Waals was able to make such a great achievement that significantly influenced this field is that he had a deep understanding of molecular behavior from the work of Boyle, Bernoulli and Clausius and the experiment of Andrew for the continuity of vapor and liquid states. These provided a basis for his ideas.

As a consequence, the concept of the internal pressure from the Van der Waals equation made an important contribution to other fields. This equation of state may be rewritten as:

$$pV = RT[1 + (b - a/kT)/V + \cdots]. \tag{2.4}$$

When $1/V$ is very small, that is, the gas is quite dilute and as a result, is not much different from ideal gas, there is no vapor-liquid phase transition in the dilute gas. If the correction of the term $1/V$ became remarkable, this equation would turn into the Van der Waals equation of state. When the temperature is quite high, the Van der Waals isotherms are close to those of an ideal gas. Therefore, what the Van der Waals equation describes is the physical behavior in the low-temperature liquid and high-density region. In contrast to the ideal gas model, the Van der Waals equation may also describe the critical phenomena.

Van der Waals' pioneering attempt greatly promoted the development of the kinetic theory for molecules and was highly recognized by the great physicist Maxwell in the nineteenth century. However, Maxwell also pointed out its deficiency. In fact, when the temperature $T \prec T_c$, there is a passage in the Van der Waals isotherms where the partial derivative $(\partial p / \partial V)_T > 0$, which is unphysical according to the basic thermodynamic principle. Maxwell pointed out that the passage where $(\partial p / \partial V)_T > 0$ in the Van der Waals isotherm corresponds to an unstable state, which should be replaced by a horizontal line, whose two ends correspond to gas and liquid states. At a given temperature, liquid and gas states may coexist in the corresponding equilibrium pressure p_s in the horizontal line, which may be determined by Maxwell's equal-area rule,

$$\int_{V_L}^{V_v} (p - p_s)dV = 0, \tag{2.5}$$

which represents how to determine the horizontal line of the isotherm across the Van der Waals loop.

The Van der Waals equation of state along with the Maxwell equal-area rule can describe the vapor-liquid phase transition well, including the critical point. It can be seen that when $T < T_c$, there is a vapor and liquid coexistence region, which is represented by the saturated vapor and liquid volumes in the ends of an isotherm. When the temperature approaches the critical point, the difference disappears, which is what was observed by Andrews (1869). It can be also noticed that the tangent line of $(\partial p / \partial V)_T$ at the critical point is parallel to the lateral axis, which means that the compressibility diverges. A very small external force can result in a large volume change, and an occasional change of pressure may lead to a relatively large fluctuation in the density.

The Van der Waals equation of state also could represent some interesting phenomena such as supercooled vapor and superheat liquid states. It was mentioned above that the passage where $(\partial p / \partial V)_T > 0$ in the Van der Waals curve does not represent an actual physical state. However, two other passages where $(\partial p / \partial V)_T > 0$ below the saturated liquid state and above the saturated vapor state do not conflict with the thermodynamic conditions and represent superheated and supercooled states, respectively.

In fact, when a gas is compressed along an isotherm, if the conditions are appropriate, it is not immediately condensed into the liquid state via the saturated vapor boundary but may be further compressed, allowing it to reach a higher isotherm while it still remains at the earlier low temperature. As a result, it becomes a supercooled gas state. Thus, it is said that the supercooled state is a metastable state. Similarly, a liquid may become a superheated liquid, another metastable state.

The Van der Waals equation of state not only qualitatively explained the basic phenomena of the vapor-liquid phase transition such as the two-phase coexistence, critical point, supercooled and superheated states, but also could be used to quantitatively predict many properties of a substance near the critical point.

2.3 Corresponding States of the Van der Waals Equation of State

At the critical point, the first and second derivatives of pressure with respect to the volume are equal to zero:

$$\left(\frac{\partial p}{\partial V}\right)_T = 0 \qquad (2.6)$$

and

$$\left(\frac{\partial^2 p}{\partial V^2}\right)_T = 0. \qquad (2.7)$$

With the above conditions and the Van der Waals equation of state, p, V, and T at the critical point could be obtained:

$$T_c = \frac{8a}{27b} \qquad p_c = \frac{a}{27b^2}, \qquad \text{and} \ \ V_c = 3b. \qquad (2.8)$$

When the three critical parameters T_c, p_c, and V_c are used as the units of temperature, pressure and volume, and three dimensionless ratios $T_r = T/T_c$, $p_r = p/p_c$, and $V_r = V/V_c$ are introduced, it is interesting that the Van der Waals equation of state became

$$\left(p_r + \frac{3}{V_r^2}\right)(3V_r - 1) = 8T_r. \qquad (2.9)$$

By means of the critical temperature, critical volume and critical pressure, Van der Waals derived the corresponding-states principle, which is not dependent on a specific substance.

Equation (2.9) does not include any substance-dependent parameter (i.e., a and b), which shows that all gases should conform to the corresponding-states principle. It was Van der Waals (1880, 1881) who derived this result, which is the first law reflecting the behavior of the phase transition universality in history (Meslin, 1893). At that time, Van der Waals also deduced some of its consequences, of which the most important is that the vapor pressure curve,

$$p/p_c = f(T_r), \qquad (2.10)$$

which should also be a universal function. Indeed, most substances conform much better to this corresponding-states principle than they do to the equation itself, if $f(T_r)$ and similar functions are used to fit the experimental results and are not determined from the equation.

The principal application of the corresponding-states principle is to predict unknown properties of many molecules from the known properties of a few. This principle was put forward too late to influence the experiments of Cailletet and Pictet in their first attempts to liquefy air, but it helped to guide later attempts on hydrogen and helium when it was realized that reasonable estimates of the critical and boiling temperatures could be made from a study of isotherms at higher temperatures, and then these could either be fit directly with the Van der Waals equation of state or matched to known gases by the corresponding-states principle. Onnes (1897) realized at once that the corresponding-states principle implied a molecular principle of similitude, although his inevitable ignorance of the form of the intermolecular potential function prevented him from formulating this quantitatively. He stated that the conditions for the principle to hold were, first, that the molecules of the different substances had to be completely hard elastic bodies of common shape; second, that the long-range forces that they exert had to emanate from corresponding points and be proportional to the same function of the corresponding separations of these; and third, that the absolute temperature had to be proportional to the mean kinetic energy of the translational motion of the molecules.

A similar idea may lie behind Gibbs' proposal for an unwritten chapter titled on similarity in thermodynamics that was found among his papers after his death in 1903 (Rowlinson, 1988). Byk (1921, 1922) deduced correctly that a fourth parameter would be needed for light molecules, which were governed by the quantum effect. Today all who face the problem of predicting physical properties would agree with the statement by Guggenheim (1945) that the corresponding-states principle could safely be regarded as the most useful derivation of the Van der Waals equation of state. The corresponding-states principle gradually became the basis to predict equilibrium properties of pure fluids and mixtures, reduced equation of state, and transport properties.

2.4 Universal Form of Corresponding-States Principle

It can be seen from Eq. (2.9) that the universal form of the corresponding-states principle is as follows:

$$F(\frac{p}{p_c},\frac{V}{V_c},\frac{T}{T_c}) = F(p_r,V_r,T_r) = 0 , \tag{2.11}$$

where $V_r = V/V_c$ and $T_r = T/T_c$. The corresponding-states principle of the equation of state is here introduced as,

$$\frac{pV}{RT} = F\left(\frac{V}{V_c}, \frac{T}{T_c}\right) = F(V_r, T_r) \tag{2.12}$$

The modern theories of the equation of state investigate the virial coefficients of the series expansion in the universal form of the corresponding-states principle,

$$\frac{pV}{RT} = 1 + B_r / V_r + C_r / V_r^2 + \cdots, \tag{2.13}$$

$$B_r = \frac{B}{V_c} = f_B(T_r) \tag{2.14}$$

and
$$C_r = \frac{C}{V_c^2} = f_C(T_r), \tag{2.15}$$

which may be considered as the basis of an understanding of the corresponding-states principle.

The value of pV/RT would be the same for all molecules if the corresponding-states principle holds rigorously. From experiments it is known that $p_c V_c / RT_c$ is only a constant for groups of gases of a similar physical and chemical behavior. The only group for which at present statistical and wave-mechanical calculations have been carried out is the group of the so-called permanent gases, such as helium, neon, argon, hydrogen, nitrogen, and oxygen. Within this group, the value $p_c V_c / RT_c$ shows a tendency to become larger for the lighter gases. As the influence of wave-mechanics is greater the lighter the molecule, it may be expected that this inconstancy of $p_c V_c / RT_c$ in this group of permanent gases can be accounted for by the wave-mechanical behavior, as was indicated by Wohl (1929).

References

Andrews, T., 1869, *Philos. Trans. Roy. Soc.* **159**: 575.
Byk, A., 1921, *Phys. Z.* **22**: 15.
Byk, A., 1921, *Ann. Phys.* **66**: 157.
Byk, A., 1922, *Ann. Phys.* **69**: 161.
Guggenheim, E. A., 1945, *J. Chem. Phys.* **13**: 253.
Meslin, G. 1893, *Compt. Rend. Acad. Sci.* **116**: 135.
Onnes, K. H., 1897, *Arch. Neerl.* **30**: 101.

Van der Waals, J. D., 1873, On the Continuity of the Gaseous and Liquid States, Rowlinson, J. S., ed. Studies in Statistical Mechanics, Vol.14 (North-Holland, Amsterdam, 1988).

Van der Waals, J. D., 1880, *Verhand. Kon. Akad. Weten. Amsterdam* **20(5)**: 1; **20(6)**: 1.

Van der Waals, J. D., 1881, *Ann. Phys.* **25**: 27.

Wohl, A., 1929, *Z. Phys. Chem.* **2B**: 77.

Chapter 3

Theoretical Basis of the Corresponding-States Principle

3.1 Two-Parameter Corresponding-States Theory of Monatomic Molecules

3.1.1 Introduction

The idea of corresponding states was suggested by Van der Waals in 1880 from the equation of state proposed by him in 1873. The theory of corresponding states is of course more general than the equation and holds that, in terms of reduced temperatures and volume, the behavior of all substances should be the same. Due to the work of London and of Pitzer on the nature of Van der Waals forces, it was found that the derivation of the theory of corresponding states was discovered on the basis of assumptions compatible with the understanding of molecular forces.

De Boer and Michels (1938) discussed the theory of corresponding states with respect to gas imperfection, but they did not consider the liquid state from this point of view. In addition to the theory of corresponding states, in terms of the critical point, there is Trouton's rule and its various modifications, most notably that of Hilderbrand (1915, 1918). While the derivation led to the Van der Waals theory of corresponding states, for certain common groups of substances it becomes equivalent to the Hilderbrand rule.

Pitzer (1939) stated a set of assumptions sufficient to lead to the Van der Waals corresponding-states theory and showed that argon, krypton, and xenon have several properties in accordance with this theory. Guggenheim (1945, 1953, 1966) further interpreted these assumptions stated then.

3.1.2 Assumptions of the Two-Parameter Corresponding-States Theory

1. Classical statistical mechanics is used. As an example, this assumption for a harmonically vibrating solid gives the Dulong and Petit value of the heat capacity.

Since most solids attain this condition before melting, this assumption should be acceptable for their liquids. However, this excludes hydrogen and helium from consideration and makes doubtful a few other cases including neon. Any distinction between Fermi-Dirac statistics and Bose-Einstein statistics is not considered. The effect of quantization of the translational degrees of freedom is negligible.

2.   It is assumed that the molecules are spherically symmetrical, either actually or by virtue of rapid and free rotation. The pair intermolecular potential function can be written as $u(r) = \varepsilon\phi(r/\sigma)$, ϕ is a universal function, r the intermolecular distance, and ε and σ are characteristic constants.

3.   The nature of any intramolecular vibrations is assumed to be the same whether the molecules are in the liquid or gas states, regardless of density.

4.   The potential energy of an assemblage of molecules is taken as a function only of the various intermolecular distances. The total potential energy for the system is the sum of all possible pairwise interactions; thus, the intermolecular potential energy has simple additivity, $u = \Sigma\phi(r)$, which is a good approximation for nonpolar molecules that have dispersion force only.

Although assumption 1 means that no real molecule fulfills assumption 4 precisely, at least the heavier rare gases relatively accurately satisfy it, from which the Van der Waals corresponding-states theory can be derived with a satisfactory degree of accuracy. The function ϕ has the inverse sixth power attraction, which is characteristic of most, if not all, molecules forming normal liquids. Furthermore, provided that the repulsive branch is very steep, minor variations in slope should not affect the results greatly. Nothing need be considered concerning the exact nature of the function ϕ ; the only assumption is its universality. In their discussion of corresponding states for gas imperfection De Boer and Michels (1938) were led to conditions essentially the same as these.

Both assumptions 1 and 2 are satisfied provided $(mkT)^{1/2} v^{1/2} \succ\succ h,$ where m is molecular mass, T absolute temperature, and v volume per molecule. As pointed out by Pitzer, this condition excludes hydrogen and helium owing to their small molecular masses and to some extent limits the applicability of the principle to neon. There is no need to assume that degrees of freedom other than the translational must be classical. In fact, the vibrational degrees of freedom of diatomic molecules are practically unexcited, and yet it will be shown that such molecules can obey the principle of corresponding states.

The effect of assumption 2 is that some diatomic and polyatomic molecules may obey the principle in the gaseous and liquid state, but cannot be expected to obey it in the solid state. Assumption 3 might be more usefully stated: "The intramolecular degrees of freedom are assumed to be completely independent of the volume per molecule". Highly polar molecules are ruled out by these assumptions. Metals and molecules capable of forming hydrogen bonds are ruled out by assumption 4.

The assumption is, however, accurate for large values of r where the intermolecular potential energy ε is proportional to $-r^{-6}$. Moreover many macroscopic properties are insensitive to the precise form of the relation between ε and r for small r. Consequently, assumption 4, though not rigorously true, turns out to be a useful approximation for many non-polar molecules.

3.1.3   Derivation of the Two-Parameter Corresponding-States Theory

Derivation of the two-parameter corresponding-states theory was first proposed by Pitzer (1939), and further described by Guggenheim (1945), by Mansoori et al. (1980), by Rowlinson and Swinton (1982), by Reed and Gubbins (1991), by Prausnitz et al. (1999), and by Ely and Marrucho (2000). If two systems fulfilling the above conditions and containing equal numbers of molecules are enclosed in containers with linear dimensions proportional to their σ's (thus volumes proportional to σ^3), it is apparent that spacial correspondence has been attained. Since energies of motion and temperature are proportional, the corresponding temperatures will be proportional to the ε/k's. Thus if the systems are compared at volumes bearing the same ratio to the values of σ^3 and at temperatures thus related to the ε/k's, then these systems should present corresponding behavior.

For the system of non-ideal gas and liquids, the interaction of particles (molecules or atoms) cannot be neglected. In general the Hamiltonian H may be thus expressed as,

$$H = \frac{1}{2m}\sum_i^N \mathbf{P}_i^2 + \Phi(\mathbf{r}^N) + \sum_i^N F(T,N), \qquad (3.1)$$

where m is the molecular mass, and F is the partition function for the internal degrees of freedom of a molecule and depends only on the temperature T. On the right hand of Eq. (3.1), the first term represents translational energy, the second is intermolecular interaction energy, and the third is internal motion (rotational and vibrational) energy. Classical statistics is considered to be applicable for the partition function Q,

$$Q = \frac{1}{h^{3N}}\frac{1}{N!}\int_V \cdots \int \exp(\frac{H}{kT})\mathrm{d}\mathbf{P}_1 \cdots \mathrm{d}\mathbf{P}_N \mathrm{d}\mathbf{r}_1 \cdots \mathrm{d}\mathbf{r}_N \qquad (3.2)$$

where h is the Planck constant, $\mathrm{d}\mathbf{r}_i = \mathrm{d}x_i \mathrm{d}y_i \mathrm{d}z_i$, and $\mathrm{d}\mathbf{P}_i = \mathrm{d}p_{xi}\mathrm{d}p_{yi}\mathrm{d}p_{zi}$. In the translational partition function, Q_{tran} is a product from the kinetic energy and from the potential energy. For a pure system of N molecules, Q_{tran} is given by:

$$Q_{tran} = (2\pi mkT / h^2)^{3N/2} Q_{conf} / N!,$$ (3.3)

where k is the Boltzmann constant, and Q_{conf} is the configurational integral that depends on temperature and volume. According to the above assumptions, Q_{conf} may be obtained as:

$$Q_{conf} = \int_V \cdots \int \exp\left[-\frac{\varepsilon}{kT} \sum_{i>j} \phi(\frac{r_{ij}}{\sigma}) \right] dr_1 \cdots dr_N .$$ (3.4)

By converting $r_1 \cdots r_N$ in Eq. (3.4) into $(r_1 / \sigma^3) \cdots (r_N / \sigma^3)$, Eq. (3.4) is:

$$Q_{conf} = \sigma^{3N} \int_{V/\sigma^3} \cdots \int \exp\left[-\frac{\varepsilon}{kT} \sum_{i>j} \phi(\frac{r_{ij}}{\sigma}) \right] d(\frac{r_1}{\sigma^3}) \cdots d(\frac{r_N}{\sigma^3}) .$$ (3.5)

The integral in Eq. (3.5) is the function of kT / ε, V / σ^3, and N, and can thus be derived as follows:

$$Q_{conf} = \sigma^{3N} \psi(V / \sigma^3, kT / \varepsilon, N) .$$ (3.6)

ψ is a universal function of kT/ε and V/σ^3, where σ corresponds to the value when the potential energy has the minimum value ε. In the infinite dilute system, ψ approaches 1. The Helmholtz free energy A is,

$$A = -kT \ln Q .$$ (3.7)

Using Eq. (3.6) and the Sterling formula, we obtain,

$$-\frac{A}{NkT} = \frac{3}{2} \ln \frac{2\pi mkT}{h^2} + \ln Q_{int}(T, N) + 3\ln \sigma + \ln \frac{\psi(V / \sigma^3, \varepsilon / kT, N)}{N} - (\ln N - 1)$$ (3.8)

If A is regarded as independent of N, then $\ln \dfrac{\psi(V / \sigma^3, \varepsilon / kT, N)}{N} - \ln N$ in Eq. (3.8) should also be independent of N. Thus we have,

$$\ln \varphi(\frac{V}{N\sigma^3}, \frac{\varepsilon}{kT}) = \ln \frac{\psi(V / \sigma^3, \varepsilon / kT, N)}{N} - \ln N$$ (3.9)

and

$$Q = \left[N\varphi(\frac{V}{N\sigma^3}, \frac{\varepsilon}{kT}) \right]^N.$$ (3.10)

Based on Eqs. (3.6) and (3.10), the equation of state may be expressed as:

$$p/kT = (\frac{\partial \ln Q}{\partial V})_{T,N} = \left[\frac{\partial \ln \varphi(V/N\sigma^3, \varepsilon/kT)}{\partial(V/N)} \right]_{T,N},$$ (3.11)

which is written as:

$$\frac{p}{\varepsilon/\sigma^3} = \frac{kT}{\varepsilon} \left[\frac{\partial \ln \varphi}{\partial(V/N\sigma^3)} \right]_{T,N}.$$ (3.12)

Eq. (3.12) thus becomes the following dimensionless generalized form:

$$p^* = f(T^*, V^*),$$ (3.13)

where f is a universal function.

$$p^* = p/(\varepsilon/\sigma^3)$$ (3.14)

$$T^* = T/(\varepsilon/k)$$ (3.15)

$$V^* = V/N\sigma^3$$ (3.16)

Eqs. (3.13) to (3.16) represent the pressure-volume-temperature relation in terms of the intermolecular potential parameters, while usually it is expressed in terms of the critical parameters p_c, T_c, and V_c from the relationships,

$$p_c^* = f(T_c^*, V_c^*)$$ (3.17)

$$(\partial p^*/\partial V^*)_c = 0$$ (3.18)

and

$$(\partial^2 p^*/\partial V^{*2})_c = 0.$$ (3.19)

When the intermolecular potential is the Lennard-Jones (6-12) equation and the derived equation of state is applied to the critical state, then we have the following relations:

$$p_c = 0.116\frac{\varepsilon}{\sigma^3}$$ (3.20)

$$V_c = \sqrt{2}\pi N \sigma^3 \tag{3.21}$$

and

$$T_c = 1.25 \frac{\varepsilon}{k}, \tag{3.22}$$

which display the corresponding-states parameters for these two states in a simple proportional relation. p_c, T_c, and V_c are substance-dependent parameters; thus Eq. (3.13) is

$$p_r = f(T_r, V_r) \tag{3.23}$$

$$p^* V^* / T^* = pV / (RT). \tag{3.24}$$

Since Eq. (3.23) and Eq. (3.24) reflect the same physical behavior, these corresponding reduced parameters should have a relation. It can be seen that Eq. (3.13) and Eq. (3.23) are the same in essence.

Described by Eqs. (3.23) and (3.13), the corresponding-states principle is actually the generalized equation of state. When the reduced parameters are used, all thermodynamic properties may be described by the universal functions.

3.1.4 Two-Parameter Corresponding-States Theory and Properties

3.1.4.1 Virial Coefficients

From the Lennard-Jones (6-12) potential function, the reduced second virial coefficient B/b is a function of kT/ε, where $b = 2N_0 \pi \sigma^3 / 3$; i.e., $2\pi V^* / 3$, kT/ε is the reduced temperature T_R.

When the critical state is used as the reduced state, the reduced second virial coefficient B_R becomes B/V_c; that is

$$B_R = B/V_c = B_r(T_r). \tag{3.25}$$

3.1.4.2 Compressibility Factor

$$Z = \frac{pV}{RT} = Z(V_R, T_R) \tag{3.26}$$

or

$$Z = Z(p_R, T_R). \tag{3.27}$$

When the critical state is used as the reduced state, the reduced compressibility factor is

$$Z = Z(p_r, T_r).$$
(3.28)

3.1.4.3 Enthalpy Departure

$$H - H^0 = -RT^2 \int_0^p (\frac{\partial Z}{\partial T})_p \, d\ln p,$$
(3.29)

where H^0 is in the reference state. Substituting the reduced parameters into the above formula,

$$\frac{H - H^0}{RT^*} = -T_R^2 \int_0^{p_R} (\frac{\partial Z(p_R, T_R)}{\partial T_R})_{p_R} \, d\ln p_R = \frac{H - H^0}{RT^*}(p_R, T_R),$$
(3.30)

where $(H - H^0)/RT^*$ is the reduced enthalpy departure, which is a universal function of p_R and T_R. When the critical state is used as the reduced state, the reduced enthalpy departure is,

$$\frac{H - H^0}{RT_c} = \frac{H - H^0}{RT_c}(p_r, T_r).$$
(3.31)

3.1.4.4 Fugacity Coefficient

The fugacity coefficient f/p is,

$$\ln(f/p) = \int_0^p (Z - 1) d\ln p$$
$$= \int_0^{p_R} [Z(p_R, T_R) - 1] d\ln p_R = \ln\frac{f}{p}(p_R, T_R)$$
(3.32)

$\ln(f/p)$ is the universal function of p_R and T_R. When the critical state is used as the reduced state, the reduced fugacity coefficient is,

$$\ln(f/p) = \ln\frac{f}{p}(p_r, T_r).$$
(3.33)

3.1.4.5 Entropy Departure

$$\frac{S-S^0}{R} = \int_0^{p_R} [1 - Z - T_R (\frac{\partial Z}{\partial T_R})_{p_R}] d \ln p_R - \ln \frac{p}{p^0}$$

$$= \frac{S-S^0}{R}(p_R, T_R) - \ln \frac{p}{p_0}$$

(3.34)

When the critical state is used as the reduced state, the reduced entropy departure is,

$$\frac{S-S^0}{R} = \frac{S-S^0}{R}(p_r, T_r) - \ln(p / p^0).$$

(3.35)

3.1.4.6 Vapor Pressure

$$p_{s,R} = p_{s,R}(T_R).$$

(3.36)

The subscript s denotes the saturated vapor repssure. As above,

$$p_{s,r} = p_{s,r}(T_r).$$

(3.37)

3.1.4.7 Viscosity

$$\eta_R = \eta_R(T_R, \rho_R).$$

(3.38)

Similarly,

$$\eta_r = \eta_r(T_r, \rho_r).$$

(3.39)

3.1.4.8 Thermal Conductivity

$$\lambda_R = \lambda_R(T_R, \rho_R).$$

(3.40)

Similarly,

$$\lambda_r = \lambda_r(T_r, \rho_r).$$

(3.41)

3.1.4.9 Surface Tension

$$\sigma_R = \sigma_R(T_R).$$

(3.42)

Similarly,

$$\sigma_r = \sigma_r(T_r).$$

(3.43)

Other properties, such as internal energy, free energy, and heat capacity departure functions, can be expressed as universal functions of p_R and T_R.

3.2 Corresponding-States Theory of Asymmetric Molecules

It has been established that the behavior of certain substances conforming to the above assumptions obeys the corresponding-states principle. Substances that deviate from the two-parameter corresponding-states principle do not completely meet the above assumptions.

Since we were unable to calculate the perfect behavior theoretically, we shall not expect to derive exact expressions for the deviation in the various cases. The discussion of these deviations is further complicated by the fact that whenever a source of imperfection exists, it usually causes all the properties of the liquid to deviate, so that none are left to specify the proper conditions for comparison. The best that can be done is to select some properties that are relatively sensitive to imperfection and insensitive to temperature, and compare these properties at points chosen by the use of properties of the opposite character.

Since the entropy of the gas depends particularly on its volume, and because imperfection cannot change the liquid volume very greatly, it seems best to compare entropies of vaporization at an equal vapor to liquid volume ratio. This basis is, moreover, a very convenient one. The change in heat capacity with vaporization is relatively insensitive to the temperature, and will be taken at the same point.

3.2.1 Derivation of Corresponding States of Asymmetric Molecules

A theoretical basis for extending the simple corresponding-states principle of Van der Waals to include fluids with nonsymmetrical intermolecular potentials can be developed for those fluids that have relatively slight deviations from spherically symmetric potentials. As noted by Leland and Chappelear (1968), the work of Cook and Rowlinson (1953) and Pople (1954) has developed a procedure for the expansion of an orientation-dependent pair potential for a symmetrical contribution in a series of spherical harmonics. This expansion as presented by Pople (1954) is:

$$u(r,\theta_1,\phi_1,\theta_2,\phi_2) = \sum_{i_1=0}\sum_{i_2=0}\sum_{m} \xi_{(r)}^{(l_1,l_2,m)} S_{l_1,m}(\theta_1,\phi_1) S_{l_2,m}(\theta_2,\phi_2). \quad (3.44)$$

The subscripts 1 and 2 represent the molecules in a pair. The θ and ϕ terms are spherical polar coordinate angles, which define the position of a central axis of symmetry through each molecule relative to a plane containing a line connecting the centers of this axis in each molecule. The center-to-center distance along this line is designated by r. The $S_{l,m}(\theta,\phi)$ functions are normalized associated Legendre polynomials for each molecule in the pair. These polynomials are functions of the angular coordinates for each molecule as follows:

$$S_{l,m}(\theta,\phi) = \left[(2l+1)\frac{(l-|m|)!}{(l+|m|)!} \right]^{1/2} P_l^{|m|}(\cos\theta)e^{im\phi}, \quad (3.45)$$

where $P_l^{|m|}$ is the associated Legendre function. The values of l are positive integers only and $|m|$ is always equal to or less than l. Physically, the integers l and m designate one particular potential configuration produced among an infinite set, which is summed to give the actual nonsymmetrical potential. The value of l governs the degree of departure from the spherical symmetry of lines of the constant potential in a plane containing the principal axis of the molecule. A value of $l = 0$ indicates that these lines of constant potential are exactly spherical. The value of m selects various degrees of orientation of the molecular axes relative to each other. Although m is summed over both positive and negative integers up to $|m| = l$, it can be shown that positive and negative values of the same m lead to identical terms.

The coefficients $\xi_{(r)}^{(l_1, l_2, m)}$ for a given pair vary only with r and involve characteristic parameters determined by the distribution of centers of attraction and repulsion within each contributing configuration. For the special case in which separation distances are large relative to the separation of force centers within each molecule, the coefficients originating from various combinations of l_1, l_2, and m have the form:

$$\xi_{(r)}^{(l_1, l_2, m)} = c(l_1, l_2, m)vr^{-n}, \tag{3.46}$$

where $c(l_1, l_2, m)$ is a constant which depends on the values of l_1, l_2, and m. The v factor is a function of the distribution of force centers in each contributing configuration.

The terms in which l_1, l_2, and m are all zero represent the symmetrical components of the total potential. Pople (1954) treats all the remaining terms as a perturbation on the symmetrical contribution to the potential. In this manner, the total potential for N molecules is:

$$U_N = \sum_{i>j} u_{ij}^0 + \sum_{i>j} u_{ij}^1, \tag{3.47}$$

where u_{ij}^0 represents the symmetrical component of the ij pair interaction and u_{ij}^1 is the total orientation effect obtained from all terms in Eq. (3.38) in which at least one of the integers l_1, l_2, or m is unequal to zero. With this separation in Eq. (3.41), the orientation portion of the potential in the configuration integral Q_c may be expanded as:

$$
Q_{conf} = \frac{1}{N!}\left(\frac{1}{4\pi}\right)^N \int \cdots 2N \cdots \int e^{-\sum_{i>j} u_{ij}^0}
$$

$$
\times \left[1 - \left(\frac{\sum_{i>j} u_{ij}^{-1}}{kT}\right) + \frac{1}{2!}\left(\frac{\sum_{i>j} u_{ij}^{-1}}{kT}\right)^2 \cdots \right](dr_1, dr_2 \cdots dr_N)(d\omega_1 d\omega_2 \cdots d\omega_N)
$$

(3.48)

where $d\omega$ represents the differential of all coordinate angles for a molecule. Pople has used this expansion of the configuration integral to determine the configurational Helmholtz free energy and the second virial coefficient. The results for the Helmholtz free energy consist of a symmetrical potential contribution $A^{(0)}$ obtained from the configuration integral using a Lennard-Jones type potential, followed by a correction $A^{(2)}$ contributed by nonsymmetrical components of the potential:

$$
A = A^{(0)} + A^{(2)} \tag{3.49}
$$

and

$$
A^{(2)} = -\left(\frac{N}{2kT}\right)[1/2\sum_{l,l',m}{}' \int [\xi_{(r)}^{(l,l',m)}]^2 n_2^{(0)}(\mathbf{r})d\mathbf{r}
$$

$$
+ \sum_{i \neq 0} \iint \xi_{(r_1)}^{(l,0,0)} \xi_{(r_2)}^{(l,0,0)} P_l(\cos\theta_{12})n_3^{(0)}(\mathbf{r_1},\mathbf{r_2})d\mathbf{r_1}d\mathbf{r_2}]
$$

(3.50)

The $\sum'$ represents the sum over all l, l', and m values for which at least one of the integers is unequal to zero. The term $n_2^{(0)}(\mathbf{r})d\mathbf{r}$ is the probable number of molecules with a symmetrical force distribution existing in the differential volume $d\mathbf{r}$ at position $\mathbf{r}$. The $n_3^{(0)}(\mathbf{r_1},\mathbf{r_2})d\mathbf{r_1}d\mathbf{r_2}$ is the probable number of molecules at $\mathbf{r_1}$ and $\mathbf{r_2}$ in a fluid of symmetrical molecules, when there is a molecule at the common origin of $\mathbf{r_1}$ and $\mathbf{r_2}$. The angle θ_{12} is the angle between the vectors $\mathbf{r_1}$ and $\mathbf{r_2}$.

Some important conclusions may be derived from these results:

1.   The $(\sum_{i>j} u_{ij}/kT)$ term in Eq. (3.48) makes no contribution at all to the thermodynamic properties. This term consists of all terms in Eq. (3.44) except the one for which l_1, l_2, and m are all zero, and these terms all vanish when integrated over all values of the coordinate angles. Consequently, the first perturbation contribution to the thermodynamic properties comes from the

$(\sum_{i>j} u_{ij}{}' / kT)^2$ term in Eq. (3.44); when this is integrated, the only terms which do not vanish are those involving $[\xi(r)]^2$ products in which l_1, l_2, and m are all identical in each $\xi(r)$ multiple.

2. Eq. (3.50) shows that when pairwise interactions are corrected for orientation effects, a new term involving three-body interactions is introduced. This term cannot be expressed entirely in terms of the two-body potentials. In the virial expansion, corrections to the third virial coefficient for nonpairwise additivity in nonsymmetric molecules will be greater than in those which are symmetric, because of the additional nonpairwise additive term originating from the orientation effects in two-body interactions.

3. For dilute gases in which the third virial can be neglected, it is possible to define a temperature-dependent pair potential for a hypothetical reference fluid obeying the simple corresponding-states principle, which can replace an actual pair potential with a small orientation dependence. The potential function for this reference has the form:

$$u(r,T) = \varepsilon(T)f[r / \sigma(T)]. \qquad (3.51)$$

A potential $u(r,\theta_1,\theta_2,\phi_1,\phi_2)$, which can be expressed as a symmetric potential plus a small perturbation, as in Eq. (3.47), can be replaced exactly by Eq. (3.51) for certain types of perturbations. From the work of Cook and Rowlinson (1953) it can be shown that, when this perturbation is caused only by a slight dipole moment, the parameters in Eq. (3.51) for a Lennard-Jones reference are:

$$\varepsilon(T) = \varepsilon_0 (1 + \frac{\mu^4}{12kT\varepsilon_0\sigma_0{}^6})^2$$

$$\sigma(T) = \sigma_0 (1 + \frac{\mu^4}{12kT\varepsilon_0\sigma_0{}^6})^{-1/6}. \qquad (3.52)$$

It is assumed in Eq. (3.52) that the molecule is spherical with a dipole at the center. Leach et al. (1967) have developed Eq. (3.51) for a perturbation due to a non-isotropic polarizability as well as a slight dipole moment, assuming that the nonsymmetrical molecular structure has no steric effects on the repulsion. The exact result for a Lennard-Jones reference is:

$$\varepsilon(T) = \frac{\varepsilon_0 (1 + \dfrac{5\mu^4}{72\varepsilon_0\sigma_0^6 kT})}{[1 - \dfrac{4\varepsilon_0\kappa^2}{kT}(1+\dfrac{7}{5}\kappa^2) - \dfrac{1}{20}(\dfrac{\mu^4\bar{\alpha}^2}{\varepsilon_0\sigma_0^{12}kT})]}$$

$$\sigma(T) = \sigma_0 \left\{ \frac{\left[1 - \frac{4\varepsilon_0 \kappa^2}{kT}(1 + \frac{7}{5}\kappa^2) - \frac{1}{20}(\frac{\mu^4 \overline{\alpha}^2}{\varepsilon_0 \sigma_0^{12} kT}) \right]}{\left(1 + \frac{5\mu^4}{72\varepsilon_0 \sigma_0^6 kT} \right)} \right\}^{1/6} . \tag{3.53}$$

The ε_0 and σ_0 parameters in Eqs. (3.52) and (3.53) are the Lennard-Jones parameters of the spherically symmetric component of the actual potential.

4. When an equation such as Eq. (3.53) can be defined to replace an orientation-dependent potential, then the pairwise additive contribution to any property of the nonsymmetric fluid may be obtained by substituting reduced parameters $[\varepsilon(T)/kT]$ and $[Nr(T)^3/V]$ into the reduced equation for this property of a fluid obeying the simple corresponding-states principle, provided that the property does not require a temperature derivative of the configuration integral in its derivation.

5. The complete expansion of Eq. (3.50) shows all the molecular parameters which may be needed for a reduced equation of state for fluids deviating from the simple corresponding-states principle. This equation is represented by:

$$Z = f(\frac{\varepsilon_0}{kT}, \frac{Nr_0^3}{V}, \frac{\mu^2}{\varepsilon_0 r_0^3}, \frac{\theta^2}{\varepsilon_0 r_0^5}, \kappa, \frac{\overline{\alpha}}{r_0^3}, \frac{\mu^2 \overline{\alpha}}{\varepsilon_0 r_0^6}, \frac{\mu^2 \theta^2}{\varepsilon_0 r_0^8} \cdots) . \tag{3.54}$$

The $\overline{\alpha}/\sigma_0^3$ term enters from an Axilrod-Teller (1943) nonadditivity correction to three-body interactions, and not directly from the perturbation treatment. The reduced $\mu^2 \overline{\alpha}$ term arises chiefly from the three-body second term, and only slightly from the perturbation.

The theoretical development is useful in understanding the nature of the corrections needed by the simple corresponding-states principle, and in providing a guide to empirical extensions. Although an equation such as Eq. (3.48) shows the variables which must be considered, its actual use is difficult because very few experimental data are available for parameters such as quadrupole moments and nonisotropic polarizabilities. Eq. (3.48) is useful in showing that the (ε_0/kT) and $(N\sigma_0^3/V)$ terms of the simple corresponding-states principle must be supplemented by reduced parameters which represent ratios of orientation-dependent forces to symmetrical forces. Ratios of this type can be defined empirically. An extended corresponding-states principle for a large group of fluids can be developed in terms of only three reduced parameters with one of the parameters defined as an empirical function of the ratio of the total contribution of orientation-dependent forces to symmetrical forces. A general reduced equation of state for fluids of this type is then:

$$Z = f(\frac{\varepsilon_0}{kT}, \frac{N\sigma_0^3}{V}, \lambda),$$ (3.55)

where λ is the empirically defined third parameter and f is a universal function for fluids. The symmetrical contribution to the total potential is characterized by the parameters ε_0 and σ_0^3. At the critical point, solving Eq.(3.49) produces,

$$Z_c = f_1(\lambda)$$ (3.56)

$$\varepsilon_0 / kT_c = f_2(\lambda)$$ (3.57)

and

$$V_c / N\sigma_0^3 = f_3(\lambda).$$ (3.58)

Parameters ε_0 and σ_0^3 are now directly proportional to the critical values multiplied by a function of the third parameter. Consequently, Equation (3.55) may also be written:

$$Z = f(T_R, V_R, \lambda).$$ (3.59)

Since one of the indications of the failure of the simple corresponding-states theory is its prediction of a universal critical compressibility factor for all fluids, it was natural to use the value Z_c as a third parameter. Su (1937) defined a third parameter (Vp_c / RT_c), which is equivalent to $(V_r Z_c)$ by Viswanath and Su (1965). The use of Z_c itself was proposed by Meissner and Seferian (1951) and was extensively used by Lydersen, Greenkorn, and Hougen (1955). Hirschfelder et al. (1958) showed that the relation between Z_c and the average or effective coordination number is affected somewhat by the molecular shape. As a third corresponding-states parameter introduced by Pitzer et al. (1955), the acentric factor has been widely used to further correlate properties (Pitzer and Curl, 1957; Curl and Pitzer, 1958; Pitzer and Hultgren, 1958; Danon and Pitzer, 1962; Lee and Kesler, 1975). A linear extension of acentric factor of the theory of Pitzer et al. to describe a larger molecule was attempted (Teja, 1979, 1980; Teja and Patel, 1981; Teja and Rice, 1981; Teja et al., 1981).

Another approach to extend the simple two-parameter corresponding-states principle was accomplished by making the intermolecular potential parameter functions of the additional characterization parameters and the thermodynamic state. This could be justified theoretically on the basis of results obtained by performing angle averaging on a non-spherical potential. The net result of this substitution is a corresponding-states model that has the same mathematical form as the simple two-parameter model, but the definitions of the dimensionless volume and temperature are more complex, as recently reviewed by Ely and Marrucho (2000) and by Huber and Hanley (1996).

3.2.2 Quantum Effects

Considering that the translational degrees of freedom are quantum, the translational partition function is,

$$Q_{tran} = \sum_{r} \exp(-E_r / kT),$$

(3.60)

where E_r is the translational energy, which can be obtained from the Schroedinger equation,

$$\left[-\frac{h^2}{8\pi^2 m} \sum_i \nabla_i^2 + E_p - E_r \right] Q = 0,$$

(3.61)

where ∇_i^2 is the Laplacian, and E_p is the potential energy of the system. If the intermolecular interaction is expressed in terms of the two characteristic parameters, the Lennard-Jones potential function is substituted into Eq. (3.61):

$$\left[-\frac{h^2}{8\pi^2 m} \sum_i \nabla_i^2 + \sum_{i>j} \varepsilon \phi(r_{ij} / \sigma) - E_r \right] Q = 0,$$

(3.62)

where ε and σ are potential parameters. Eq. (3.62) can be written in the reduced form:

$$\left[-\frac{h^2}{8\pi^2 m \varepsilon \sigma^2} \sum_i \nabla_{Ri}^2 + \sum_{i>j} \phi(r_{ij} / \sigma) - E_r / \varepsilon \right] Q = 0.$$ (3.63)

∇_{Ri}^2 is the reduced Laplacian:

$$\nabla_{Ri}^2 \equiv \frac{\partial^2}{\partial(x_i / \sigma)^2} + \frac{\partial^2}{\partial(y_i / \sigma)^2} + \frac{\partial^2}{\partial(z_i / \sigma)^2}.$$

(3.6

4)

In Eq. (3.63), the reduced E_r / ε is dependent not only on the number of molecules N and V/σ, but also on the quantum parameter $\Lambda_R = h/\sigma\sqrt{m\varepsilon}$. As in a similar derivation to the classical corresponding-states principle, we have,

$$p_R = f(T_R, V_R, \Lambda_R). \tag{3.65}$$

Consequently, Eq. (3.65) is the generalized expression of the corresponding-states principle, in which the quantum effect is considered. Uhlenbeck and Beth (1936) and Kirkwood (1933) have shown that for high temperatures, the second virial coefficient B, after the introduction of quantum mechanics in the theory of the equation of state, can be expressed in a series expansion in powers of $\hbar^2$, where $\hbar = h/2\pi$:

$$B = B_{vdW} + B^{(1)} + B^{(2)}. \tag{3.66}$$

B_{vdW} was the classical expression and $B^{(1)}$ and $B^{(2)}$ were given by De Boer and Michels (1938) as,

$$B_{vdW} = -2\pi N \int \left(e^{-\phi(r)/kT} - 1\right) r^2 \mathrm{d}r \tag{3.67}$$

$$B^{(1)} = 2\pi N \frac{\hbar^2}{12mk^3T^3} \int e^{-\phi(r)/kT} \left(\frac{\mathrm{d}\phi}{\mathrm{d}r}\right)^2 r^2 \mathrm{d}r \tag{3.68}$$

$$B^{(2)} = 2\pi N \frac{\hbar^4}{120m^2k^4T^4} \int e^{-\phi(r)/kT}$$
$$\left\{\left(\frac{\mathrm{d}^2\phi}{\mathrm{d}r}\right)^2 + \frac{2}{r^2}\left(\frac{\mathrm{d}\phi}{\mathrm{d}r}\right)^2 + \frac{10}{9rkT}\left(\frac{\mathrm{d}\phi}{\mathrm{d}r}\right)^3 - \frac{5}{36k^2T^2}\left(\frac{\mathrm{d}\phi}{\mathrm{d}r}\right)^4\right\} r^2 \mathrm{d}r. \tag{3.69}$$

The integration can be carried out after introduction of the Lennard-Jones potential function for $\phi(r)$, and expanding $\exp\left(4\varepsilon/kTR^6\right)$ into a series. The integral is then developed in a series of Γ-functions resulting in the expression:

$$B_{vdW} = \frac{2\pi N\sigma^3}{3}\left(\frac{4\varepsilon}{kT}\right)^{1/4} \sum_{i=0}^{\infty} c_i \left(\frac{4\varepsilon}{kT}\right)^{i/2}, \tag{3.70}$$

in which $c_i = \frac{-1}{4i!}\Gamma\left(-\frac{1}{4} + \frac{i}{2}\right)$;

$$B^{(1)} = \frac{2\pi N\sigma^3}{3} \frac{h^2}{m\varepsilon\sigma^2} \left(\frac{4\varepsilon}{kT}\right)^{13/12} \sum_{i=0}^{\infty} c_i^{(1)} \left(\frac{4\varepsilon}{kT}\right)^{i/2}, \qquad (3.71)$$

in which $c_i^{(1)} = \frac{36i - 11}{768\pi^2 i!}\Gamma\left(-\frac{1}{12} + \frac{i}{2}\right);$

$$B^{(2)} = \frac{2\pi N\sigma^3}{3} \left(\frac{h^2}{m\varepsilon\sigma^2}\right)^2 \left(\frac{4\varepsilon}{kT}\right)^{23/12} \sum_{i=0}^{\infty} c_i^{(2)} \left(\frac{4\varepsilon}{kT}\right)^{i/2}, \qquad (3.72)$$

in which $c_i^{(2)} = \frac{3024i^2 + 4728i + 767}{491520\pi^4 i!}\Gamma\left(\frac{1}{12} + \frac{i}{2}\right).$

The values of the coefficients are given in Table 3.1

Table 3.1 Values of Coefficients c_i

i	$10^3 c_i^{(1)}$	$10^6 c_i^{(2)}$	i	$10^6 c_i^{(1)}$	$10^6 c_i^{(2)}$
0	18.84	184.5	7	19.18	2.75
1	6.80	317.0	8	4.92	806
2	4.26	447.0	9	1.171	217
3	1.89	130.6	10	0.269	55.0
4	0.709	61.6	11	0.0586	13.15
5	0.234	24.6	12	0.01215	2.98
6	0.0698	6.86	13	0.002415	0.645

Hence, substituting $\Lambda = h/\sqrt{m\varepsilon}$ and introducing the molecular units of volume and temperature $2\pi N\sigma^3/3$ and ε/k, the expression for $B*$ of the corresponding-states principle can be written:

$$B^* = \frac{B}{2\pi N\sigma^3/3} = (\frac{4}{T^*})^{1/4} \sum_0^{\infty} c_i (\frac{4}{T^*})^{i/2} + (\frac{\Lambda}{\sigma})^2 (\frac{4}{T^*})^{13/12} \sum_0^{\infty} c_i^{(1)} (\frac{4}{T^*})^{i/2}$$
$$+ (\frac{\Lambda}{\sigma})^4 (\frac{4}{T^*})^{23/12} \sum_0^{\infty} c_i^{(2)} (\frac{4}{T^*})^{i/2} + \cdots \qquad (3.73)$$

$B*$ is thus not only a function of the reduced temperature but depends also on the value of Λ/σ and $T^* = T/(\varepsilon/k)$. It is therefore of interest to investigate the physical meaning of $\Lambda = h/\sqrt{m\varepsilon}$. The wavelength of the De Broglie wave of a molecule moving with momentum p is $\lambda = h/p$ and the De Broglie wavelength of relative motion of two molecules is equal to $\lambda = h/\sqrt{m\varepsilon}$. Λ can therefore be interpreted as the De Broglie wavelength of relative motion at a temperature at which the mean relative kinetic energy is equal to ε. If it is assumed that ε/kT, Λ/σ can be interpreted as the reduced De Broglie wavelength of relative motion at reduced temperature.

It is now possible to give a physical interpretation of the parameter introduced by Byk (1921, 1922): $\zeta = \left(h/\sqrt{mkT_c}\right)/\left(V_c/N\right)^{1/3}$ and $\Lambda_c = h/\sqrt{mkT_c}$ are the mean De Broglie wavelength of relative motion of the molecules at the critical temperature. Therefore, ζ can be given the same interpretation of the reduced De Broglie wavelength at the critical temperature.

The failure of neon to exactly follow the two-parameter corresponding-states theory was already ascribed to quantum-mechanical effects. The deviation of hydrogen and helium are even more marked. As pointed out by Byk (1921, 1922), ζ can be used to measure the extent of quantum deviations considering the critical point as the reference state.

In many respects, the nature of a liquid and solid are similar. It has been commonly assumed in treating liquids that most of the time a molecule vibrates within a cage of neighboring molecules in about the same manner as in the solid. Thus, if the ability of molecules in the liquid to occasionally jump from cage to cage is of no particular significance to the property of interest, the difference between liquid and solid may be slight. With this idea in mind, we take the well-known Debye functions for a somewhat idealized solid, and apply them to the liquid. The changes in C_p and S are in opposite directions. The heat capacity becomes too small and the entropy of the liquid becomes too large. This leaves the entropy of vaporization too small.

The liquid volume is also affected similarly. Quantum-mechanical effects cause the volume to remain too large and the coefficient of expansion to become too small. This, of course, changes the gas to liquid volume ratio, which is the standard for comparing entropies. The liquid volume of neon is about 3% too large at its triple point, and this change affects the entropy by less than 0.4 J/K, which is almost negligible. The experimental entropy of vaporization of neon deviates from the perfect by 2.8 J/K as compared to the calculated 4.2 J/K, which is probably as good agreement as can be expected.

References

Axilrod, B. M. and E. Teller, 1943, *J. Chem. Phys.* **11**: 299.
Byk, A., 1921, *Phys. Z.* **22**: 15.
Byk, A., 1921, *Ann. Phys.* **66**: 167.

Byk, A., 1922, *Ann. Phys.* **69**: 161.

Cook, D. and J. S. Rowlinson, 1953, *Proc. Roy. Soc. A* **219**: 405.

Curl, R. F., Jr. and K. S. Pitzer, 1958, *Ind. Eng. Chem.* **50**: 265.

Danon, F. and K. S. Pitzer, 1962, *J. Chem. Phys.* **36**: 425; *J. Phys. Chem.* **66**: 583.

De Boer, J. and A. Michels, 1938, *Physica* **5**: 945.

Ely, J. F. and I. M. F. Marrucho, 2000, in *Equations of State for Fluids and Fluid Mixtures*, J. V. Sengers, R. F. Kayser, C. J. Peters, and H. J. White Jr. eds. (Elsevier, Amsterdam).

Guggenheim, E. A., 1945, *J. Chem. Phys.* **13**: 253.

Guggenheim, E. A., 1953, *Rev. Pure Appl. Chem.* **3**: 1.

Guggenheim, E. A., 1966, Applications of Statistical Mechanics (Clarendon Press, Oxford).

Hildebrand, H., 1915, *J. Am. Chem. Soc.* **37**: 970.

Hildebrand, H., 1918, *J. Am. Chem. Soc.* **40**: 45.

Hirschfelder, J. O., R. J. Buehler, H. A. McGee, and J. R. Sutton, 1958, *Ind. Eng. Chem.* **50**: 375, 386.

Kirkwood, J. G., 1933, *Phys. Rev.* **44**: 31.

Huber, M. L. and H. J. Hanley, 1996, in Transport Properties of Fluids: Their Correlation, Prediction and Estimation, J. Millat, J. H. Dymond, and C. A. Nieto de Castro, eds. (Cambridge University Press, New York).

Leach, J. W., 1967, Molecular Structure Corrections for Application of the Theory of Corresponding States to Nonspherical Pure Fluids and Mixtures, PhD thesis, Rice University, Houston, Tex.

Lee, B. I. and M. G. Kesler, 1975, *AIChE J.* **21**: 510, 1040, 1237.

Leland, T. W. and P. S. Chappelear, 1968, *Ind. Eng. Chem.* **60(7)**: 15.

Lydersen, A. L., R. A. Greenkorn, and O. A. Hougen, 1955, University of Wisconsin Eng. Sta. Rept. 4.

Mansoori, G. A., V. Patel, and M. Edalat, 1980, *Int. J. Thermophys.* **1**: 285.

Meissner, H. P. and R. Seferian, 1951, *Chem. Eng. Prog.* **47**: 579.

Pitzer, K. S., 1939, *J. Chem. Phys.* **7**: 583.

Pitzer, K. S., 1955, *J. Am. Chem. Soc.* **77**: 3427.

Pitzer, K. S. and R. F. Curl, Jr., 1957, *J. Am. Chem. Soc.* **79**: 2369.

Pitzer, K. S. and G. O. Hultgren, 1958, *J. Am. Chem. Soc.* **80**: 4793.

Pitzer, K. S., D. Z. Lippmann, R. F. Curl, Jr., C. M. Huggins, and D. E. Petersen, 1955, *J. Am. Chem. Soc.* **77**: 3433.

Pople, J. A., 1954, *Proc. Roy. Soc. A* **221**: 498.

Prausnitz, J. M., R. N. Lichtenthaler, and E. G. de Azevedo, 1999, Molecular Theory of Fluid-Phase Equilibria, 3rd ed. (Prentice Hall, New Jersey).

Reed, T. M. and K. E. Gubbins, 1991, Applied Statistical Mechanics: Thermodynamic and Transport Properties of Fluids (Butterworth-Heinemann, Boston).

Rowlinson, J. S. and F. L. Swinton, 1982, Liquids and Liquid Mixtures, 3rd ed. (Butterworths, London).

Su, G. S., D. Sc. Thesis, 1937, M.I.T., Cambridge, Mass.

Teja, A. S., 1979, *Ind. Eng. Chem. Fundam.* **18**: 435.

Teja, A. S., 1980, *AIChE J.* **26**: 337.

Teja, A. S. and N. C. Patel, 1981, *Chem.Eng.Commun.* **13**: 39.

Teja, A. S. and P. Rice, 1981, *Ind. Eng. Chem. Fundam.* **20**: 77; *Chem. Eng. Sci.* **36**: 1.

Teja, A. S., S. I. Sandler, and N. C. Patel, 1981, *Chem. Eng. J.* **21**: 21.

Uhlenbeck, G. E. and E. Beth, 1936, *Physica* **3**: 729.

Viswanath, D. S. and G. J. Su, 1965, *AIChE J.* **11**: 202.

Chapter 4

The Corresponding-States Parameters

4.1 Critical Parameters

The quantitative representation of the behavior of thermodynamic properties of fluids over both gas and liquid regions has proven to be an unusually difficult problem. It is 125 years since Van der Waals discovered the corresponding-states principle. Our understanding of the forces operating between molecules has developed rapidly since the emergence of a satisfactory quantum theory in 1926. Particularly important is the work of London on the attractive forces between molecules. This situation is in marked contrast to the extensive theoretical advances with respect to the thermodynamic properties of ideal gases.

The Riedel parameter, which is similar to the acentric factor, was applied to those molecules excluding highly polar or associated liquids. Rowlinson also presented certain theoretical work, which yielded a similar correlation scheme.

The most generally useful method of prediction of the volumetric properties of fluids is the hypothesis of corresponding states, which also came originally from Van der Waals (1873, 1880, 1881). Engineers have used this method extensively to obtain estimated properties for design purposes, and many authors have presented charts of both volumetric and related thermodynamic properties on the basis of corresponding states (Pitzer, 1955).

As described in Sections 2.3 and 2.4, the two-parameter corresponding-states principle in terms of the critical parameters is

$$\frac{p}{p_c} = f(\frac{T}{T_c}, \frac{V}{V_c}),\tag{4.1}$$

which is only applicable to monatomic molecules and may cause about 30% error when describing normal fluids.

4.2 Acentric Factor

The difference between spherical molecules and nonspherical molecules was investigated by Pitzer (1955). The greater the deviation from sphericity, the more difficult for two molecules to approach each other. As a result, the potential minimum of nonspherical molecules is narrower than that of spherical molecules. In the liquid state, this decreases the probability of any given clustering of nonspherical molecules compared with spherical molecules. Likewise, decreased probabilities lead to decreased entropies. Thus, a comparison of the entropies of nonspherical molecules (subscript noq) and of spherical molecules (subscript q) in the liquid state is obtained:

$$S_{l,q} \succ S_{l,noq} .$$
(4.2)

Thus, the decrease in entropy from the ideal gas state to some comparable state in the imperfect gas or liquid region is greater for nonspherical molecules than for spherical molecules. Then, for the entropies of vaporization $\Delta S = S_v - S_l$ for spherical and nonspherical molecules we have,

$$\Delta S_{l,noq} \succ \Delta S_{l,q} .$$
(4.3)

The relation of the entropy of vaporization to the vapor pressure is,

$$\frac{\partial \ln p_r}{\partial (1/T_r)} = -\frac{T_r \Delta S}{R} .$$
(4.4)

Based on Eqs. (4.3) and (4.4), at the same temperature, the slope of the vapor pressure curve of the spherical molecules is smaller than nonsphercial molecules:

$$\left. \frac{\partial \ln p_r}{\partial (1/T_r)} \right|_q \prec \left. \frac{\partial \ln p_r}{\partial (1/T_r)} \right|_{non,q} .$$
(4.5)

The larger the departure of a nonspherical molecule from the spherical shape, the more the increase in slope of the vapor-pressure curve. For the simple fluids, such as argon, krypton, and xenon, the reduced vapor pressure is almost precisely 0.1 at a reduced temperature of 0.7, as found by Pitzer et al. (1955). This point is well removed from the critical point yet above the melting point for almost all substances. Consequently, Pitzer et al. (1955) defined the acentric factor in terms of the reduced vapor pressure at this reference value of the reduced temperature:

$$\omega = -\log p_r \big|_{T_r = 0.7} - 1 ,$$
(4.6)

where $p_r = p/p_c$, $T_r = T/T_c$. It can be seen that ω is a corresponding-states parameter to measure the difference from spherical molecules. The slope of the vapor pressure curve is closely related to the entropy of vaporization. The acentric factor can be regarded as a measure of the increase in the entropy of vaporization over that of spherical molecules. The acentric factor depends upon the core radius of a globular molecule, the length of an elongated molecule, or the dipole moment of a slightly polar molecule. The corresponding-states theory of Pitzer et al. (1955) is thus as follows:

$$\frac{p}{p_c} = f(\frac{T}{T_c}, \frac{V}{V_c}, \omega) . \tag{4.7}$$

4.3 Aspherical Factor

The maximum error in predicted volume is about 15%, while errors in other predicted properties range from 5% to 45 %, from the Van der Waals corresponding-state theory for nonpolar normal molecules. These percentages indicate the extent to which actual substances deviate from the Van der Waals corresponding-states theory. The corresponding-states theory of Pitzer et al. yields results for nonpolar substances about one order of magnitude more accurate than Van der Waals'. Only critical parameters and the acentric factor of each substance are required in the corresponding-states theory of Pitzer et al., however, highly polar molecules, for example, ammonia and water, did not conform to the corresponding-states theory of Pitzer et al.

For polar molecules, the interaction potential of permanent dipoles of fixed orientation varies as the inverse third power of the distance between their centers. Therefore, a cluster of non-rotating dipolar molecules have properties significantly different from those of nonpolar molecules (Pitzer, 1955). The physical concept of the fourth parameter for polar molecules corresponds to an intermolecular potential function that accounts for dispersion and dipole-dipole interaction effects to describe typical polar molecules. The dipole and quadrupole moments influence the physical behavior in the critical state, which includes critical-universality behaviors. Since the critical temperature characterizes the intermolecular interaction energy while the critical volume is related to the intermolecular distance, the critical state represents the essential physical behavior of a molecule. The critical compression factor $Z_c = p_c/\rho_c RT_c$, where R is the gas constant, rapidly decreases for highly nonspherical molecules and may reflect the deviation of the behavior of a highly nonspherical molecule relative to that of monatomic molecules. The Van der Waals corresponding-states theory gives the same value of Z_c for all substances. In fact, values of Z_c range from 0.2 to

0.29:

$$Z_{c,q} \succ Z_{c,noq}. \tag{4.8}$$

Hirschfelder et al. (1958) showed that the relation between Z_c and the average or effective coordination number is affected somewhat by the molecular shape. The expression $Z_c = 0.29 - 0.08\omega$ describes the relation between the critical compression factor and the acentric factor for nonpolar molecules (Pitzer et al., 1955) rather accurately, but it does not apply to highly nonspherical molecules. To extend the quadratic expression for the acentric factor of the corresponding-states theory of Pitzer et al., Eq. (4.7), to polar molecules, a fourth parameter θ can be introduced (Xiang, 1998, 2000, 2001, 2002, 2003):

$$\theta = (Z_c - 0.29)^2, \tag{4.9}$$

where $Z_c = 0.29$ is the critical compression factor for monatomic molecules. The extended corresponding-states parameter θ is defined as a new physical term called the *aspherical factor*, which reflects the deviation of the behavior of a highly nonspherical molecule relative to the spherical molecule argon, which is similar to the definition of the acentric factor with Z_c as a characteristic parameter for a specific substance. Consequently, the corresponding-states principle is expressed as,

$$\frac{p}{p_c} = f(\frac{T}{T_c}, \frac{V}{V_c}, \omega, \theta), \tag{4.10}$$

where p represents the pressure from the Van der Waals corresponding-states theory for spherical symmetry [Eq. (4.23)], for weakly nonspherical molecules [Eq. (4.7)], and for highly nonspherical molecules [Eq. (4.10)]. The physical significance of the extended corresponding-states parameter θ differs from the acentric factor ω. The approximation $\theta_{nonpolar} = 0.0064\omega^2$ is valid for nonpolar substances. The extended corresponding-states parameter θ significantly reflects the behavior of highly polar molecules. For example, the critical compression factor of water calculated from this expression is 0.263, which seems to indicate a nonpolar molecule, but the actual value is 0.228 for this strongly hydrogen-bonding molecule. The ratios of $\theta/\theta_{nonpolar}$ vary up to 5 as shown in Fig. 4.1 for some representative highly polar molecules, which results in an error in properties such as the vapor pressure of up to 45 % as shown in Fig. 4.2. The data in Fig. 4.1 show that even small alcohols such as ethanol and propanol behave more like heavy hydrocarbons as found by Wilding et al. (1987).

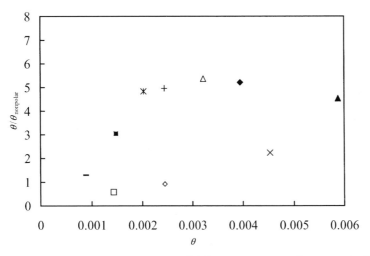

Fig. 4. 1 Comparison of the ratios of $\theta / \theta_{nonpolar}$, where $\theta_{nonpolar} = 0.0064\omega^2$, with the quadratic expression for the acentric factor of the corresponding-states theory of Pitzer et al. (+) Difluoromethane; (-) 1,1,1,2-Tetrafluoroethane; (■) 1,1-Difluoroethane; (*) Ammonia; (♦) Water; (×) Methanol; (▲) Acetic acid; (Δ) Acetone; (◊) Ethanol; (□) Propanol. [Reproduced with permission from Xiang (2003)]

The dependence of the pressure on the temperature and volume is generally expressed as the compression factor $Z = pV / RT$, which is also considered to obey the corresponding-states principle (Xiang, 2003):

$$Z = pV / RT = f(\frac{T}{T_c}, \frac{V}{V_c}, \omega, \theta). \tag{4.11}$$

The critical compression factor for a substance is obtained from Eq. (4.9) at the critical point as,

$$Z_c = 0.29 + c_1\omega + c_2\theta, \tag{4.12}$$

where the general parameters c_1 and c_2 are –0.055 and –11.5 for the weak and highly nonspherical contributions to the deviation of a substance from the value for the spherical molecule for argon $Z_c = 0.29$. The value $c_1 = -0.08$ for normal molecules was found by Pitzer et al. (1955). Eq. (4.12) may be arranged as,

$$c_2\theta + (0.29 - Z_c) + c_1\omega = 0. \tag{4.13}$$

Substituting Eq. (4.9), Eq. (4.13) can be written,

$$c_2 Z_c^2 - (1 + 2 \times 0.29 c_2) Z_c + (0.29^2 c_2 + 0.29 + c_1 \omega) = 0 . \qquad (4.14)$$

This can be seen to describe the compression factor for the deviation of a real nonspherical molecule from the spherical molecule for argon. Eq. (4.14) shows that the extended corresponding-states theory, Eq. (4.11), is a quadratic expression for Z_c at the critical point; in contrast, the corresponding-states theory of Pitzer et al., Eq. (4.7), is a linear expression for Z_c at the critical point:

$$-Z_c + 0.29 + c_1 \omega = 0 . \qquad (4.15)$$

The nature of the acentric and aspherical parameters makes it possible to separate the effect of the molecular structure of nonspherical nonpolar and polar molecules from that of spherical molecules. The corresponding-states parameters, acentric factor and aspherical factor, account for the behavior of highly nonspherical molecules to represent the properties of all classes of molecules. Together with general arguments, this result has indicated that this extended corresponding-states theory might well yield a complete description for properties of all classes of molecules, which is at least one order of magnitude more accurate than that obtained from the corresponding-states theory of Pitzer et al. This is illustrated for the case of vapor pressure, in Figs. 4.2 and 4.3.

4.3 Properties from the Corresponding-States Principle

4.4.1 Second and Third Virial Coefficients

$$B_r = B_r^{(0)}(T_r) + \omega B_r^{(1)}(T_r) + \theta B_r^{(2)}(T_r) \qquad (4.15)$$

$$C_r = C_r^{(0)}(T_r) + \omega C_r^{(1)}(T_r) + \theta C_r^{(2)}(T_r) \qquad (4.16)$$

4.4.2 Compressibility Factor

$$Z = Z^{(0)}(p_r, T_r) + \omega Z^{(1)}(p_r, T_r) + \theta Z^{(2)}(p_r, T_r) \qquad (4.17)$$

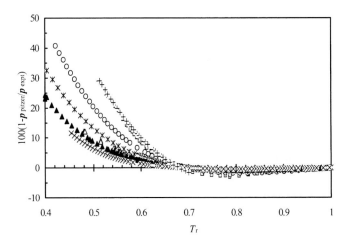

Fig. 4.2 Comparison of the vapor-pressure data with values calculated from the corresponding-states theory of Pitzer et al. The errors show that the predictions of the corresponding-states principle for a strongly nonspherical molecule have errors up to 45% for low temperatures and up to 5% for high temperatures. (*) Difluoromethane; (×) 1,1,1,2-Tetrafluoroethane; (▲) 1,1-Difluoroethane; (Δ) Ammonia; (○) Water; (□) Methanol; (+) Acetic acid. [Reproduced with permission from Xiang (2003)]

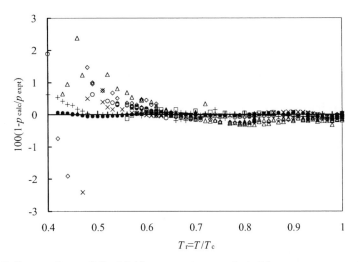

Fig 4.3 Comparison of the highly accurate experimental vapor-pressure data with values calculated from the extended corresponding-states theory. (□) Argon; (Δ) Ethane; (◊) Propane; (×) 1,1,1,2-Tetrafluoroethane; (○) 1,1-Difluoroethane; (+) Difluoromethane; (●) Water. [Adapted from Xiang (2001a)]

4.4.3 Enthalpy Departure

$$\frac{H-H^0}{RT_C} = \left.\frac{H-H^0}{RT_C}\right|^{(0)}(p_r,T_r) + \omega\left.\frac{H-H^0}{RT_C}\right|^{(1)}(p_r,T_r) + \theta\left.\frac{H-H^0}{RT_C}\right|^{(2)}(p_r,T_r) \quad (4.18)$$

4.4.4 Fugacity

$$\ln\varphi = \ln\varphi^{(0)}(p_r,T_r) + \omega\ln\varphi^{(1)}(p_r,T_r) + \theta\ln\varphi^{(2)}(p_r,T_r) \quad (4.19)$$

4.4.5 Entropy Departure

$$\frac{S-S^0}{R} + \ln(p/p^0) = \left.\frac{S-S^0}{R}\right|^{(0)}(p_r,T_r) + \omega\left.\frac{S-S^0}{R}\right|^{(1)}(p_r,T_r) + \theta\left.\frac{S-S^0}{R}\right|^{(2)}(p_r,T_r)$$

$$(4.20)$$

4.4.6 Vapor Pressure

$$p_{s,r}(T_r) = p_{s,r}^{(0)}(T_r) + \omega p_{s,r}^{(1)}(T_r) + \theta p_{s,r}^{(2)}(T_r) \quad (4.21)$$

4.4.7 Density

$$\rho_r(T_r) = \rho_r^{(0)}(T_r) + \omega\rho_r^{(1)}(T_r) + \theta\rho_r^{(2)}(T_r) \quad (4.22)$$

4.4.8 Heat Capacity

$$\frac{C_p-C_p^0}{R} = \left.\frac{C_p-C_p^0}{R}\right|^{(0)}(p_r,T_r) + \omega\left.\frac{C_p-C_p^0}{R}\right|^{(1)}(p_r,T_r) + \theta\left.\frac{C_p-C_p^0}{R}\right|^{(2)}(p_r,T_r)$$

$$(4.23)$$

4.4.9 Viscosity

$$\frac{\eta}{\eta^0} = \left.\frac{\eta}{\eta^0}\right|^{(0)}(\rho_r,T_r) + \omega\left.\frac{\eta}{\eta^0}\right|^{(1)}(\rho_r,T_r) + \theta\left.\frac{\eta}{\eta^0}\right|^{(2)}(\rho_r,T_r) \quad (4.24)$$

4.4.10 Thermal Conductivity

$$\frac{\lambda}{\lambda^0} = \frac{\lambda}{\lambda^0}\bigg|^{(0)} (\rho_r, T_r) + \omega \frac{\lambda}{\lambda^0}\bigg|^{(1)} (\rho_r, T_r) + \theta \frac{\lambda}{\lambda^0}\bigg|^{(2)} (\rho_r, T_r) \qquad (4.25)$$

4.4.11 Surface Tension

$$\frac{\sigma}{\sigma^0} = \frac{\sigma}{\sigma^0}\bigg|^{(0)} (\rho_r, T_r) + \omega \frac{\sigma}{\sigma^0}\bigg|^{(1)} (\rho_r, T_r) + \theta \frac{\sigma}{\sigma^0}\bigg|^{(2)} (\rho_r, T_r) \qquad (4.26)$$

References

Hirschfelder, J. O., R. J. Buehler, H. A. McGee, and J. R. Sutton, 1958, *Ind. Eng. Chem.* **50**: 375, 386.

Pitzer, K. S., 1955, *J. Am. Chem. Soc.* **77**: 3427.

Pitzer, K. S., D. Z. Lippmann, R. F. Curl, C. M. Huggins, and D. E. Petersen, 1955, *J. Am. Chem. Soc.* **77**: 3433.

Van der Waals, J. D., 1873, On the Continuity of the Gaseous and Liquid States, J. S. Rowlinson ed. Studies in Statistical Mechanics, 1988, Vol.14 (North-Holland, Amsterdam).

Van der Waals, J. D., 1880, *Verhand. Kon. Akad. Weten. Amsterdam* **20(5)**: 1; **20(6)**: 1.

Van der Waals, J. D., 1881, *Ann. Phys. Beibl.* **5**: 27.

Wilding, W. V., J. K. Johnson, and R. L. Rowley, 1987, *Int. J. Thermophys.* **8**: 717.

Xiang, H. W., 1998, A new simple extended corresponding-states principle for polar fluids: vapor pressure and second virial coefficient, in Proc. 15th IUPAC Int. Conf. Chem. Thermodyn. (Porto, Portugal).

Xiang, H. W., 2000, Properties and the corresponding-states principle, in Proc. AIChE Annual Meeting (Los Angeles, USA).

Xiang, H. W., 2001a, *Int. J. Thermophys.* **22**: 919.

Xiang, H. W., 2001b, *J. Phys. Chem. Ref. Data* **30**: 1161.

Xiang, H. W., 2001c, *Fluid Phase Equilib.* **187**: 221.

Xiang, H. W., 2002, *Chem. Eng. Sci.* **57**: 1439.

Xiang, H. W., 2003, Thermophysicochemical Properties of Fluids: Corresponding -states Principle and Practice (Science Press, Beijing), in Chinese.

The Corresponding-States Practice

Chapter 5

Thermodynamic Properties

5.1 Thermodynamic Relations

5.1.1 Basic Thermodynamic Functions

The relations to describe the Helmholtz and Gibbs energies, enthalpies, entropies, heat capacity, speed of sound, and fugacity coefficients are presented. These relations are then used with equations of state for the calculation of enthalpy and entropy departure functions and fugacity coefficient.

Thermodynamic properties at p and T are compared with values marked with a supercript 0, corresponding to the same temperature but in an ideal-gas state and at a reference pressure p^0. In the reference state at T and p^0, the molar volume V^0 is given by:

$$V^0 = RT / p^0.$$
(5.1)

As shown below, the departure functions of a thermodynamic property can be expressed in terms of the PVT properties of a fluid. Generally, the PVT properties of a fluid are characterized by an equation of state explicit in pressure. When temperature and pressure are the independent variables, the compressibility factor is expressed as $Z = f(T_r, p_r)$.

For an equation of state with an explicit pressure, the departure function for the Helmholtz energy A, and from this result all other departure functions, can readily be obtained via the thermodynamic relation:

$$A - A^0 = -\int_\infty^V \left(p - \frac{RT}{V} \right) dV - RT \ln \frac{V}{V^0}.$$
(5.2)

In Eq. (5.2), the departure function for A depends upon the choice of V^0. Other departure functions are readily obtained from Eq. (5.2):

$$S - S^0 = \frac{-\partial}{\partial T}\left(A - A^0\right)_V , \tag{5.3}$$

and then

$$S - S^0 = \int_\infty^V \left[\left(\frac{\partial p}{\partial T}\right)_V - \frac{R}{V}\right]dV + R\ln\frac{V}{V^0} , \tag{5.4}$$

$$H - H^0 = \left(A - A^0\right) + T\left(S - S^0\right) + RT(Z - 1), \tag{5.5}$$

$$U - U^0 = \left(A - A^0\right) + T\left(S - S^0\right), \tag{5.6}$$

$$G - G^0 = \left(A - A^0\right) + RT(Z - 1). \tag{5.7}$$

Also, the fugacity coefficient can be expressed similarly:

$$\ln\frac{f}{p} = \frac{A - A^0}{RT} + \ln\frac{V}{V^0} + (Z - 1) - \ln Z$$

$$= -\frac{1}{RT}\int_\infty^V\left(P - \frac{RT}{V}\right)dV + (Z - 1) - \ln Z , \tag{5.8}$$

where $Z = pV/RT$ is the compressibility factor.

Therefore, from any pressure-explicit equation of state and a definition of the reference state (p^0 or V^0), all departure functions can readily be obtained.

An alternate calculation path to obtain departure functions is more convenient if the equation of state has an explicit volume, or if pressure and temperature are the independent variables. In such cases, an ideal gas at the same temperature as that of the system is chosen as a reference state. The reference pressure is p^0, and Eq. (5.1) applies. With the analog to the Helmholtz energy Eq. (5.2), the Gibbs energy and consequently the other thermodynamic properties can also be obtained and the departure functions can be evaluated with PVT data and a defined reference state. Generally, an equation of state incorporating the corresponding-states principle may be used to characterize pVT behavior.

5.1.2 Overview of Equations of State

5.1.2.1 Ideal Gas Properties and Development of Statistical Mechanics

In 1662, Boyle introduced the concept of pressure to obtain a gas law from experiments and qualitatively interpret properties of air. Newton thought that the repulsion force between a gas molecule and the surrounding particles made gas pressure vary inversely to the volume. Combination of the Charles law and

Gay-Lussac law with the Avogadro assumption led to the discovery of the ideal gas law of $pV = RT$, where p is the pressure, V is the volume, R is the gas constant, and T is the temperature. Subsequently, it has been found that this relation was not applicable to some states of a substance. As a consequence, some scientific constraint was given, that is, the above gas law is applicable only to the so-called ideal gas. So what is the ideal gas? We can only respond that it is the gas whose compressibility factor is close to 1 within its tolerable error. Thus, the ideal gas model appeared. This model helped to understand the thermodynamic properties of the state of substance. The Van der Waals, virial and other equations of state quantitatively described the deviation of an actual fluid state from the ideal-gas state. The extension of the ideal-gas law not only promoted the continuing refinement of the molecular theory , but also, more importantly, provided the basic idea of a method for extending its application to other properties. In particular, the corresponding-states principle that derived from the Van der Waals equation of state is on what this book focuses. It is thermodynamics that plays an important role in the study of physical properties of substances. Thermodynamics is a branch of thermophysics and chemistry that is intimately related to and interdependent with the molecular theory. For example, the Clausius-Clapeyron equation provides a useful method for obtaining enthalpies of vaporization from more easily measured vapor-pressure data. The concept of chemical potential derived from the second law of thermodynamics is basic to an understanding of chemical and phase equilibria, and the Maxwell's equations provide ways to obtain important thermodynamic properties of a substance from pressure-volume-temperature relations.

Maxwell started in 1859 and completed in 1866 the Maxwellian distribution of molecular velocities in the kinetic theory of gases. The statistical mechanics maintained a rapid development from the pioneer work of Ludwig Boltzmann and Josiah Willard Gibbs. By means of the statistical mechanics, the statistical laws of molecular system were understood, the quantitative relation of thermodynamic and transport properties with molecules and molecular velocity was established, and the thermodynamic equations of a series of thermodynamic functions were formed. All these promoted thermodynamics become a general scientific subject with rigorous system and flexible application. Disagreement between the properties calculated from the kinetic theory of gases and experimental results led to the interaction behavior between molecules being investigated. It is basic for the study of molecular behavior that the molecules attract each other at large separation distances, whereas repulsion occurs at small distances. The Lennard-Jones potential and other potentials quantitatively described such a behavior of the attraction and repulsion between molecules. More recently, some potential functions considered the effects of molecular shape and asymmetric charge distribution in the polar molecules.

5.1.2.2 Cubic Equation of State

Redlich and Kwong (1949) proposed a modified vdW-type equation of state, the so-called Redlich-Kwong equation of state, which is able to describe properties of non-polar, slightly polar and even certain polar real gases. It is applicable for densities up to slightly higher than half the critical density. The Redlich-Kwong equation of state has been used to more accurately represent vapor-liquid equilibrium and thermodynamic properties in the critical region. An improvement in the description of phase-equilibrium thermodynamic properties was given later by use of the equations of state proposed by Soave (1972) and by Peng and Robinson (1976). However, these equations of state are not applicable to strong polar fluids. The cubic equation of state was studied extensively by Fuller (1976), by Harmens and Knapp (1980), by Schmidt and Wenzel (1980), by Kubic (1982), by Patel and Teja (1982), by Peneloux et al. (1982), by Adachi et al. (1983, 1984), by Watson et al. (1986), by Trebble and Bishnoi (1987), by Yu and Lu (1987), by Iwai et al. (1988), and by Guo and Du (1989). In principle, these modified cubic equations of state do not constitute a breakthrough and are unable to overcome the limitation of the Van der Waals equation of state in describing behavior of fluids.

5.1.2.3 Multiparameter Equations of State

During the development of the cubic equations of state, multi-parameter equations of state for real gases were proposed that were based on an understanding of the Van der Waals intermolecular forces. In 1901, the virial equation of state based on statistical mechanics was proposed (Hirschfelder et al., 1954). Coefficients in the virial equation, namely the virial coefficients, could be expressed by the relation involving the intermolecular potential energy function as established from statistical mechanics. However, to date, the equation of state has not been calculated theoretically because the limitations of present knowledge. As a result, the virial equation of state is only applicable to a limited region of low densities. Beattie and Bridgemann (1928) proposed a five-parameter equation of state for real gases, whose coefficients were determined by fitting experimental data. Benedict et al. (1940) added the density terms to the Beattie-Bridgemann equation of state to describe the high-density region and obtained an 8-parameter equation of state applicable to the vapor, liquid and saturation region for hydrocarbon, nonpolar and slightly polar gases at reduced densities of less than 1.8. Because this equation of state could reproduce experimental data well in its applicable range, it was extensively recognized and improved. This equation of state, and its modifications, was generalized to apply to many types of compounds. Starling (1972) modified this equation of state to be an 11-parameter equation of state, which could describe the PVT properties of substances at reduced temperatures as low as 0.3 and reduced densities of about 3. Jacobsen and Stewart

(1973) proposed a 32-term modified BWR equation of state to describe a wide range of *pVT* and other thermodynamic properties of oxygen and nitrogen, but this modified equation of state was not applicable to the critical region. Lee and Kesler (1975) obtained the Lee-Kesler equation of state from the corresponding-states BWR equations of state. Subsequently, Teja et al. (1981), Wu and Stiel (1985), and Liu and Xu (1991) improved the corresponding-states Lee-Kesler equation of state.

5.1.2.4 Martin-Hou Equation of State

Some analytic equations of state were proposed that were based on the physical behavior on the thermodynamic surface, such as the Martin-Hou equation of state (Martin and Hou, 1955). The Martin-Hou equation of state, whose coefficients could be determined more simply than other similar equations of state, could describe the pressure-density-temperature relation of hydrocarbons and polar gases. In 1981, the applicable range of this equation of state was extended to the liquid phase (Hou et al., 1981). Subsequently, some physical meaning of the Martin-Hou equation of state from statistical mechanics was elucidated (Zhang and Hou, 1987, 1989).

5.1.2.5 Hard Sphere Equations of State

Carnahan and Starling (1969,1972) proposed a hard-sphere equation of state that was based on statistical mechanics. De Santis et al. (1976) modified this equation and obtained the Carnahan-Starling-De Santis equation of state. Subsequently, the Carnahan-Starling-De Santis equation of state was improved by Ishikawa et al. (1980), Guo et al. (1985), Kubic (1986), Mulla and Yesavage (1989), Li et al. (1991), Li and Xu (1993) and others.

5.1.2.6 Nonanalytic Equations of State

Wilson (1971, 1983) related the scaling laws to the renormalization group to enable the scaling laws to be converted into differential equations. He developed a method to divide the problem into a sequence of simpler problems, in which each part can be solved individually. In 1972, Wilson and Fisher proposed the expansion of critical exponents. As a result, critical phenomena have been verified extensively experimentally; in particular, critical exponents were proved to be independent from systems. Wilson was awarded the Nobel Prize for physics due to his preeminent work on the interpretation of physical behavior of critical phenomena. Wilson's renormalization-group theory indicates that thermodynamic properties of substances show asymptotic singular scaling behavior near the critical point and conform to universal critical exponents and universal scaling functions. However, the validity of the asymptotic power laws is restricted to a very small region of temperature and density near the critical point. Therefore, accurate experimental data near the critical point are required to determine system-dependent coefficients in the asymptotic scaling laws. Asymptotic scaling thermodynamic laws may be obtained from the critical theory of Landau, Ginzburg, and Wilson. The Landau-Ginzburg-Wilson theory was used to describe thermodynamic behavior in the critical region by Nicoll and Bhattacharjee (1981), by Nicoll and Albright (1985), by Bagnuls and Bervillier (1985), by Bagnuls et al. (1987), by Albright et al. (1986), by Dohm (1987), by Schloms and Dohm (1989), and by Chen et al. (1990). Nicoll (1981) and Chen et al. (1990) transformed a classical Landau expansion to incorporate the effects of critical fluctuations to represent the nonasymptotic behavior of fluids including the crossover from the Ising behavior in the immediate vicinity of the critical point to classical behavior far away from the critical point. Luettmer-Strathmann et al. (1992) proposed a parameter crossover model to attempt to incorporate the description of singular asymptotic behavior to classical behavior. Kiselev et al. (1988, 1991) attempted to phenomenologically propose a crossover model for the classical behavior far away from the critical point. Based on the work of Anisimov et al. (1992), Kiselev and Sengers (1993) proposed a phenomenological crossover model, which can be used to describe thermodynamic properties in the critical region, including temperatures up to 2 times the critical temperature and a limited saturated region below the critical point.

5.2 Virial Equation of State

5.2.1 Introduction

The virial equation of state is a polynomial series in the density, and is explicit in pressure and can be derived from statistical mechanics:

$$p = RT\rho + RT\rho^2 B + RT\rho^3 C + \cdots \qquad (5.9)$$

The parameters B, C, ... are called the second, third, ... virial coefficients and are functions only of temperature for a pure fluid. One reason for the equation's popularity is that the coefficients B, C, ... can be related to parameters chatacterizing the intermolecular potential function. Little accurate experimental information and calculated values using realistic potential functions are available for the third and higher virial coefficients. The virial equation is also truncated to contain only the second and third virial coefficients as follows:

$$Z = 1 + B\rho + C\rho^2 . \qquad (5.10)$$

Equation (5.10) should not be used if the density $\rho \geq \rho_c / 2$, where higher virial coefficients are required.

5.2.2 Virial Coefficients of Lennard–Jones Molecules

It is possible to account for the behavior of the second virial coefficient B as a function of temperature if the Lennard–Jones potential is written in the form:

$$\varphi(r) = \frac{\lambda}{r^n} - \frac{\mu}{r^6}. \tag{5.11}$$

The values of n are between 9 and 14. In general, the value $n = 12$ is used.
Based on Eq. (5.11), we have:

$$\varphi(r) = 4\varepsilon \left(\frac{1}{R^{12}} - \frac{1}{R^6} \right) \tag{5.12}$$

$$\varepsilon = (\mu^2 / 4\lambda), \quad \sigma = (\lambda / \mu)^{1/6}, \tag{5.13}$$

where $R = r / \sigma$. Introducing Eq. (5.12) into the known theoretical expression for B:

$$B = -2\pi N \int_0^\infty \left[\exp(-\phi(r)/kT) - 1 \right] r^2 dr. \tag{5.14}$$

For the Lennard–Jones potential, the second virial coefficient is expressed as:

$$B = \frac{2\pi N \sigma^3}{3} (\frac{4\varepsilon}{kT})^{1/4} \sum_{i=0}^\infty c_i (\frac{4\varepsilon}{kT})^{i/2}, \tag{5.15}$$

where

$$c_i = \frac{-1}{4i!} \Gamma \left(-\frac{1}{4} + \frac{i}{2} \right). \tag{5.16}$$

Taking $2\pi N \sigma^3 / 3$ as the unit of volume and ε / k as the temperature, the reduced form for the virial equation of state is

$$\frac{pV}{RT} = 1 + \frac{B^*}{V^*} + \frac{C^*}{V^{*2}} + \cdots, \tag{5.17}$$

where $V^* = 3V / 2\pi N \sigma^3$, $T^* = kT / \varepsilon$,

$$B^* = \frac{B}{2\pi N\sigma^3 / 3} = \left(\frac{4}{T^*}\right)^{1/4} \sum c_i \left(\frac{4}{T^*}\right)^{i/2},$$ (5.18)

and

$$C^* = C/(2\pi N\sigma^3 / 3)^2.$$ (5.19)

This theory was based on the assumption that the molecular field is spherical and can be expressed by Eq. (5.12) containing two parameters ε and σ.

5.2.3 Second Virial Coefficients of Real Molecules

Dymond and Smith (1980) compiled the experimental data for the second and third virial coefficients. Most of the predictive methods are based on the integral of the intermolecualr potential via Eq. (5.14). However, since methods to determine the intermolecular potential are limited, the prediction of virial coefficients from the corresponding-states principle has been used instead.

Pitzer and Curl (1957) proposed an expression for the second virial coefficient of nonpolar molecules, which is a function of reduced temperature and the acentric factor as follows:

$$\frac{Bp_c}{RT_c} = f^{(0)}(T_r) + \omega f^{(1)}(T_r)$$ (5.20)

$$f^{(0)}(T_r) = 0.1445 - 0.330/T_r - 0.1385/T_r^2 - 0.0121/T_r^3$$ (5.21)

$$f^{(1)}(T_r) = 0.073 - 0.46/T_r - 0.50/T_r^2 - 0.097/T_r^3 - 0.0073/T_r^8,$$ (5.22)

where $f^{(0)}(T_r)$ is fitted from the experimental data for the simple molecules of argon, krypton, and xenon, and $f^{(1)}(T_r)$ is obtained from data for nonpolar molecules whose acentric factors have a relatively high value.

A similar relation was proposed by O'Connell and Prausnitz (1967) for the second virial coefficient, which is inferior to Eqs. (5.20) to (5.22) to some extent, especially in their reliability of calculation. Kreglewski (1969) and Polak and Lu (1972) also correlated the second virial coefficient for polar substances. Their method requires one or more parameters to be determined from experimental data; otherwise, the calculated error would be too large to be acceptable.

Tsonopoulos (1974) proposed a modified Pitzer-Curl correlation applicable to nonpolar, polar, nonpolar and slightly hydrogen-bonding substances. For polar substances, two substance-dependent parameters a and b were introduced as follows:

$$\frac{Bp_c}{RT_c} = f^{(0)}(T_r) + \omega f^{(1)}(T_r) + f^{(2)}(T_r) \tag{5.23}$$

$$f^{(0)}(T_r) = f^{(0)}(T_r) - 0.000607/T_r^8 \tag{5.24}$$

$$f^{(1)}(T_r) = 0.0637 + 0.331/T_r^2 - 0.423/T_r^3 - 0.008/T_r^8 \tag{5.25}$$

$$f^{(2)}(T_r) = a/T_r^6 - b/T_r^8. \tag{5.26}$$

Tsonopoulos (1975) further calculated the second virial coefficient of some polar associating substances. The Pitzer-Curl equation and the Tsonopoulos equation accurately represented the second virial coefficient of normal molecules but were not good for polar substances. For example, Weber (1994) found that by deleting the last term in the Tsonopoulos equation, a better representation of the behavior of halocarbon substances was obtained.

Hayden and O'Connell (1975) proposed a correlation for the second virial coefficient based on the intermolecular potential and required the critical temperature, critical pressure, mean radius of gyration, dipole moment, and an additional parameter to describe chemical association. Tarakad and Danner (1977) proposed a correlation for the second virial coefficient for polar molecules based on the corresponding-states principle, which required the critical temperature, critical pressure, radius of gyration, and a parameter determined from the experimental data for the second virial coefficient.

Based on the expansion of the second virial coefficients from the Lennard-Jones potential funcion, a simple empirical equation with the least number of adjustable parameters is used (Xiang, 2002, 2003):

$$B_r = B\rho_c = -b_0 T_r^{-3/4} \exp(b_1 T_r^{-3}) + b_2 T_r^{-1/2}, \tag{5.27}$$

where b_0, b_1, and b_2 are substance-dependent parameters. The extended corresponding-states principle describes the second virial coefficient as follows:

$$b_0 = b_{00} + b_{01}\omega + b_{02}\theta,$$
$$b_1 = b_{10} + b_{11}\omega + b_{12}\theta,$$
$$b_2 = b_{20} + b_{21}\omega + b_{22}\theta. \tag{5.28}$$

The critical parameters and the acentric factors of these substances are given in Table 5.1. The experimental data as given in Table 5.2 are the most accurate to date and have been widely used to test proposed models. Most critical parameters are taken from the recommended data (Ambrose and Tsonopoulos, 1995; Gude and Teja, 1995). The general coefficients of Eq. (5.28) are listed in Table 5.3. For the second virial coefficient, although b_{02} and b_{22} in Eq. (5.28) were found to be zero, they may have small values to better represent experimental data when future measurement accuracy is improved.

Figures 5.1 to 5.5 show how the extended corresponding-states theory represents recent second-virial-coefficient data for various typical real fluids over their entire temperature ranges. In the figures, only one symbol is used for all of the data sets for each fluid for the sake of clarity. For most cases the agreement between the model and the data is generally comparable to the agreement between the different data sets for each substance. Generally, the extended corresponding-states theory can describe the second virial coefficient within the experimental uncertainties over their entire temperature ranges.

In the low-temperature region, the second virial coefficient is subject to large uncertainties especially at very low temperatures. Usually, the experimental uncertainties for the second virial coefficient can be 5% to 10% at low temperatures. In general, the experimental difficulties in obtaining accurate low temperature data mean that experimental results are scarce and uncertainties are large, and the data are even less accurate at very low temperatures near the triple-point temperature. The results in Fig. 5.1 show that the predictions for the second virial coefficient at the lowest temperatures deviate by 5% even for argon for which highly accurate experimental data are available. As shown in Fig 5.2, the relative deviation is 8% at the lowest temperature 290 K (T_r=0.515) and 5% at 310 K (T_r=0.55) for benzene, which has a triple-point temperature of 278.7 K (T_r=0.496). Considering that a very small low-temperature vapor-phase compressibility change has a large effect on the second virial coefficient, the new method may predict the vapor-phase compressibility for any class of fluids over the whole temperature range within experimental uncertainty for the compressibility.

In contrast to existing models, the extended corresponding-states principle can describe the second virial coefficient of highly polar and hydrogen-bonding and associating substances, such as dichlorodifluoromethane, 1,1,1,2-tetrafluoroethane, 1,1-difluoroethane, water, methanol, ammonia, acetone, and acetonitrile as shown in Figs. 5.3 to 5.5. As Weber (1994) stated, the reduced dipole moment cannot follow the strongly negative second virial coefficient behavior of the data with differences ranging from 20% at 450 K (T_r=0.825) to almost a factor of two at 300 K (T_r=0.55) for acetonitrile. However, the extended corresponding-states principle accurately predicts the second virial coefficient of acetonitrile as shown in Fig. 5.5. The reduced dipole moment cannot be used as a fourth parameter since the dipole moment location and direction with respect to the entire molecule has an important influence on the compressibility. Neither can the reduced dipole moment be used to describe the second virial coefficients of less polar fluids (O'Connell and Prausnitz, 1967; Tsonopoulos, 1974, 1975; Weber, 1994; Tsonopoulos and Dymond, 1997), but the extended corresponding-states theory gives excellent results, as shown in Fig. 5.3, for example, for dichlorodifluoromethane and 1,1-dichloro-2,2,2-trifluoroethane, which have smaller polar dipole moments.

Table 5.1 Molar Mass M**, Critical Temperature** T_c**, Critical Pressure** p_c**, Critical Density** ρ_c**, Acentric Factor** ω**, Critical Compression Factor** z_c **and Aspherical Factor** θ

Substance	M kg/kmol	T_c (K)	p_c (kPa)	ρ_c (kg.m^{-3})	ω	$\theta \times 10^3$
Argon	39.948	150.69	4863	535	0	0.000
Nitrogen	28.013	126.19	3395	313	0.037	0.000
Methane	16.043	190.564	4598	162	0.011	0.007
Carbon dioxide	44.010	304.13	7377	468	0.225	0.245
Ethylene	28.054	282.35	5041	215	0.086	0.096
Ethane	30.070	305.32	4872	206	0.099	0.097
Propane	44.097	369.83	4248	220	0.152	0.171
Benzene	78.114	562.05	4895	305	0.210	0.472
Decane	142.86	617.70	2110	228	0.489	1.061
R12	120.91	384.95	4100	560	0.178	0.180
R22	86.470	369.28	4988	525	0.221	0.501
R32	52.020	351.26	5780	430	0.277	2.560
R123	152.93	456.83	3662	550	0.282	0.481
R125	120.02	339.17	3620	570	0.306	0.388
R134a	102.03	374.18	4055	511	0.327	0.885
R152a	66.050	386.41	4516	369	0.275	1.474
Diethyl ether	74.123	466.7	3640	258	0.281	0.420
Ammonia	17.031	405.4	11345	225	0.256	2.055
Water	18.015	647.10	22050	325	0.344	3.947
Methanol	32.042	512.0	8000	270	0.560	4.486
Ethanol	46.069	514.0	6140	276	0.644	2.514
1-Propanol	60.096	536.8	5170	274	0.623	1.292
Acetone	58.080	508.1	4700	270	0.308	2.569
Acetic acid	60.052	593.0	5786	330	0.450	5.844
Acetonitrile	41.053	545.4	4835	225	0.338	9.151

Table 5.2. Data Sources of the Second Virial Coefficient B of Substances Listed in Table 5.1

Substance	Refs.
Argon	Dymond and Smith, 1980
Nitrogen	Dymond and Smith, 1980
Methane	Dymond, 1986
Carbon dioxide	Dymond and Smith, 1980
Ethylene	Dymond and Smith, 1980
Ethane	Dymond, 1986
Propane	Dymond, 1986
Benzene	Dymond and Smith, 1980
R12	Natour et al., 1989; Kohlen et al., 1985
R22	Demiriz et al., 1993; Natour et al., 1989; Kohlen et al., 1985
R32	Bignell and Dunlop, 1993
R123	Goodwin and Moldover, 1991
R125	Boyes and Weber, 1995; Bignell and Dunlop, 1993
R134a	Tillner-Roth and Baehr, 1992; Goodwin and Moldover, 1990
R152a	Bignell and Dunlop, 1993; Tillner-Roth and Baehr, 1992
Diethyl ether	Dymond and Smith, 1980
Ammonia	Dymond and Smith, 1980
Water	Eubank et al., 1988; Dymond and Smith, 1980
Methanol	Boyes et al., 1992; Kerl and Varchmin, 1991; Abusleme and Vera, 1989; Olf et al., 1989; Tsonopoulos et al., 1989; Dymond and Smith, 1980
Ethanol	Abusleme and Vera, 1989; Tsonopoulos et al., 1989; Dymond and Smith, 1980
1-Propanol	Tsonopoulos et al., 1989; Dymond and Smith, 1980
Acetone	Dymond and Smith, 1980
Acetic acid	Dymond and Smith, 1980
Acetonitrile	Demiriz et al., 1993; Olf et al., 1989; Dymond and Smith, 1980

Table 5.3. General Coefficients of Eqs.(5.28)

b_{00}	4.553	b_{10}	0.02644	b_{20}	3.530
b_{01}	4.172	b_{11}	0.075	b_{21}	4.297
b_{02}	0	b_{12}	16.5	b_{22}	0

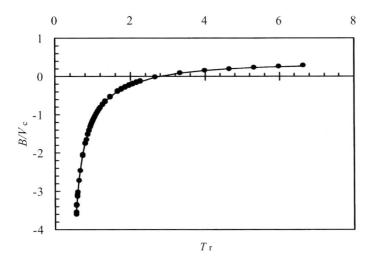

Fig. 5.1 Comparison of second-virial-coefficient data for argon with values calculated from the extended corresponding-states method, Eq. (5.27). [Reproduced with permission from Xiang (2002)]

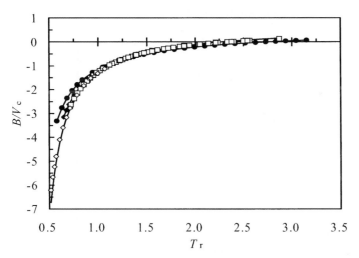

Fig. 5.2 Comparison of second-virial-coefficient data for normal substances with values calculated from the new simple extended corresponding-states method, Eq. (5.27). ($\bullet$) Methane; ($\square$) Carbon dioxide; (Δ) Propane; ($\blacktriangle$) Ethylene; ($\lozenge$) Benzene. [Reproduced with permission from Xiang (2002)]

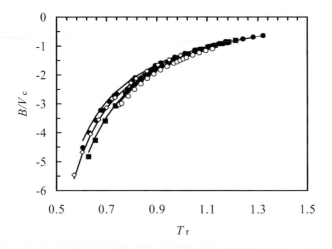

Fig. 5.3. Comparison of second-virial-coefficient data for chlorofluoro- and hydrofluoro-carbons with values calculated from the extended corresponding-states method, Eq. (5.27). (●) Chlorodifluoromethane; (◊) 1,1-Dichloro-2,2,2-trifluoroethane; (■) 1,1,1,2-Tetrafluoroethane; (○) 1,1-Difluoroethane. [Reproduced with permission from Xiang (2002)]

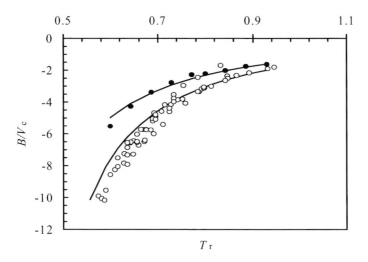

Fig. 5.4 Comparison of second-virial-coefficient data for diethyl ether and acetone with values calculated from the new simple extended corresponding-states method, Eq. (5.27). (●) Diethyl ether; (○) Acetone. [Reproduced with permission from Xiang (2002)]

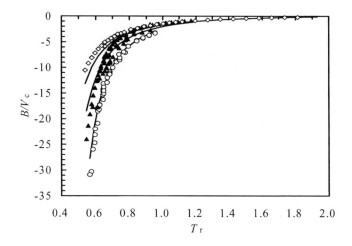

Fig. 5.5 Comparison of second-virial-coefficient data for water, methanol, and acetonitrile with values calculated from the new simple extended corresponding-states method, Eq. (5.27). (◊) Water; (▲) Methanol; (○) Acetonitrile. [Reproduced with permission from Xiang (2002)]

5.2.4 Departure Functions of the Virial Equation of State

Departure functions for the virial equation of state truncated in the second virial coefficient are:

$$z = 1 + \frac{Bp}{RT}, \tag{5.29}$$

$$A - A^0 = RT \ln \frac{V}{V - B} - RT \ln \frac{V}{V^0}, \tag{5.30}$$

and

$$S - S^0 = -\frac{RT}{V - B} \frac{dB}{dT} - R \ln \frac{V}{V - B} + R \ln \frac{V}{V^0}. \tag{5.31}$$

5.3 Crossover Equation of State

5.3.1 Crossover Equation of State

Chen et al. (1990) introduced a six-term Landau expansion for the Helmholtz free energy into a renormalized thermodynamic free energy that includes the effects of critical fluctuations and the crossover from Ising behavior in the immediate vicinity of the critical point to classical behavior far away from the critical point.

To define a reduced density difference and a reduced temperature difference,

$$\Delta\tilde{\rho} = \tilde{\rho} - 1, \quad \Delta\tilde{T} = \tilde{T} - 1. \tag{5.32}$$

The Helmholtz free energy density may be then decomposed as follows:

$$\tilde{A} = \Delta\tilde{A} + \tilde{\rho}\tilde{\mu}_0\left(\Delta\tilde{T}\right) + \tilde{A}_0\left(\Delta\tilde{T}\right). \tag{5.33}$$

$\tilde{\mu} = \left(\partial\tilde{A}/\partial\tilde{\rho}\right)_{\tilde{T}}$ and $\tilde{\chi} = (\partial^2\Delta\tilde{A}/\partial\Delta\tilde{\rho}^2)^{-1}_{\Delta\tilde{T}}$, in which $\tilde{\chi}$ is directly related to the long-range fluctuations near the critical point. The analytic background functions $\tilde{A}_0\left(\Delta\tilde{T}\right)$ and $\tilde{\mu}_0\left(\Delta\tilde{T}\right)$ are represented by truncated Taylor expansions:

$$\tilde{A}_0\left(\Delta\tilde{T}\right) = -1 + \sum_{k=1}^{4} \tilde{A}_k\left(\Delta\tilde{T}\right)^k \tag{5.34}$$

$$\tilde{\mu}_0(\Delta\tilde{T}) = \sum_{k=0}^{3} \tilde{\mu}_k(\Delta\tilde{T}_k) \tag{5.35}$$

$\Delta\tilde{A}$ in Eq. (5.33) contains all the effects of critical fluctuations that are related to $\tilde{\chi}^{-1} = (\partial^2\Delta\tilde{A}/\partial\Delta\tilde{\rho}^2)_{\Delta\tilde{T}}$.

In the classical mean-field theory, $\Delta\tilde{A}$ can be represented by a Landau expansion of the form

$$\Delta\tilde{A}_{c1} = \frac{1}{2}tM^2 + \frac{a_{04}}{4!}M^4 + \frac{a_{05}}{5!}M^5 + \frac{a_{06}}{6!}M^6 + \frac{a_{14}}{4!}tM^4 + \frac{a_{22}}{2!2!}t^2M^2 + \ldots\ldots, \tag{5.36}$$

where a_{04}, a_{05}, a_{06}, a_{14}, and a_{22} are system-dependent coefficients. The temperature-like variable t and density-like order parameter M are related to the physical variables $\Delta\tilde{T}$ and $\Delta\tilde{\rho}$ through a transformation to be specified below.

Incorporating the effects of critical fluctuations, a crossover Helmholtz-energy density $\Delta\tilde{A}_x$ can be constructed as:

$$\Delta\tilde{A}_x = \frac{1}{2}tM^2\,TD + \frac{u^*\bar{u}\Lambda}{4!}M^4\,D^2U + \frac{1}{5!}a_{05}M^5\,D^{5/2}VU + \frac{1}{6!}a_{06}M^6\,D^3U^{3/2}$$

$$+ \frac{1}{4!}a_{14}tM^4\,TD^2U^{1/2} + \frac{1}{2!2!}a_{22}t^2M^2\,T^2DU^{1/2} - \frac{1}{2}t^2k^2 \tag{5.37}$$

Here T, D, U, V and K are rescaling functions, which are themselves defined in terms of a crossover function Y:

$$T = Y^{(2v-1)/\Delta s}, \qquad D = Y^{-\eta v/\Delta s}, \qquad U = Y^{v/\Delta s}$$

$$V = Y^{(\Delta_A-1/2)/\Delta s}, \quad k = \frac{v}{c\bar{u}\Lambda}(Y^{-\alpha/\Delta s} - 1), \tag{5.38}$$

with u^* representing a universal fixed-point coupling constant. η, v, and $\alpha = 2 - 3v$ are the critical exponents. Δ_A and Δ_S are the critical exponents of the first symmetric and asymmetric correction terms to the asymptotic power laws, respectively. The values of these exponents are given as follows:

$$u^* = 0.472, \quad \eta = 0.03333, \quad v = 0.630, \quad \Delta_S = 0.51, \text{ and } \Delta_A = 1.32 \tag{5.39}$$

which are the values from three-dimensional Ising systems. The crossover parameters $\bar{u}$ and Λ in Eqs. (5.37) and (5.38) are system-dependent coefficients, which reflect the discrete microscopic structure of a substance.

The crossover function Y, which appears in expression (5.38), is determined from a set of coupled algebraic equations

$$1 - (1 - \bar{u})Y = \bar{u}Y^{v/\Delta s}\left[1 + \left(\frac{\Lambda}{\kappa}\right)^2\right]^{1/2} \tag{5.40}$$

and

$$\kappa^2 = tT + \frac{u^*\bar{u}\Lambda}{2}M^2 + DU + \frac{1}{3!}a_{05}M^3D^{3/2}VU + \frac{1}{4!}a_{06}M^4D^2U^{3/2}$$

$$+ \frac{1}{2}a_{14}tM^2TDU^{1/2} + \frac{1}{2}a_{22}t^2T^2U^{-1/2} \tag{5.41}$$

Eq. (5.41) defines the function κ that serves as a measure of distance from the critical point.

The singular asymptotic critical behavior is recovered as Y approaches 0 and Λ/κ tends to infinity, and the mean-field limit as Y approaches 1 and $\Lambda/\kappa \to 0$.

A mixing transformation of the variables in the Landau expansion t and M related to $\Delta\tilde{T}$ and $\Delta\tilde{\rho}$ is given as,

$$t = c_t \Delta \widetilde{T} + c(\partial \Delta \widetilde{A}_x / \partial M)_t \tag{5.42}$$

$$M = c_\rho (\Delta \widetilde{\rho} - d_1 \Delta \widetilde{T}) + c(\partial \Delta \widetilde{A}_x / \partial M)_t (\partial \Delta \widetilde{A}_x / \partial t)_M, \tag{5.43}$$

where c_t, c_ρ, c and d_1 are system-dependent coefficients. Thus, the expression for $\Delta \widetilde{A}$ becomes

$$\Delta \widetilde{A} = \Delta \widetilde{A}_x - c \left(\frac{\partial \Delta \widetilde{A}_x}{\partial M} \right)_t \left(\frac{\partial \Delta \widetilde{A}_x}{\partial t} \right)_M. \tag{5.44}$$

c_t and c_ρ determine the scale of the physical variables $\Delta \widetilde{T}$ and $\Delta \widetilde{\rho}$ with respect to the theoretical variables t and M, while c and d_1 arise because of vapor-liquid asymmetry near the critical point in the Ising mode.

The thermodynamic properties may be calculated from the Helmholtz free-energy density as functions of temperature and density. The crossover parameters $\bar{u}$ and Λ; the scaling-field parameters c_t, c_ρ, c and d_1, a_{ij} of the Landau expansion; and the background parameter $A_k (k \geq 1)$ and $\widetilde{\mu}_k (k \geq 2)$ are determined by fitting the crossover model to experimental data.

5.3.2 Corresponding States of the Crossover Equation of State

These parameters in the above crossover equation of state for some typical substances (Edison and Sengers, 1999; Jin et al., 1992; Van Pelt and Sengers, 1995; Tang et al., 1991; Wyczalkowska and Sengers, 1999; Wyczalkowska et al., 2000) are determined from the extended corresponding-states theory as follows (Xiang, 2003):

$$\bar{u} = 0.246 + 0.919\omega - 57.1\theta, \tag{5.45}$$

$$\Lambda = 1.27 - 2.05\omega + 481\theta, \tag{5.46}$$

$$c_t = 1.24 + 1.53\omega + 196\theta, \tag{5.47}$$

$$c_\rho = 2.63 - 0.744\omega - 75.8\theta, \tag{5.48}$$

$$c = -0.0422 - 0.0918\omega - 19.1\theta, \tag{5.49}$$

$$d_1 = -0.408 - 1.48\omega + 54.6\theta, \tag{5.50}$$

$$a_{05} = 0.0463 + 0.412\omega + 14.2\theta, \tag{5.51}$$

$$a_{06} = 0.119 + 2.21\omega - 14.5\theta, \tag{5.52}$$

$$a_{14} = 0.612 - 0.105\omega - 72.1\theta, \tag{5.53}$$

$$a_{22} = 1.13 + 8.36\omega - 163\theta. \tag{5.54}$$

5.4 Cubic Equations of State

5.4.1 Generalized Cubic Equations of State

A cubic equation of state implies an equation, which, when expanded, would contain volume terms raised to the first, second, and third powers. Many of the common two-parameter cubic equations can be expressed by the equation

$$p = \frac{RT}{V-b} - \frac{a}{V^2 + ubV + wb^2}. \tag{5.55}$$

An equivalent form of Eq. (5.55) is

$$Z^3 - (1 + B^* - uB^*)Z^2 + (A^* + wB^{*2} - uB^* - uB^{*2})Z - A^*B^* - wB^{*2} - wB^{*3} = 0, \tag{5.56}$$

where

$$A^* = \frac{ap}{R^2T^2} \tag{5.57}$$

and

$$B^* = \frac{bp}{RT}. \tag{5.58}$$

Four well-known cubic equations are the Van der Waals, Redich-Kwong (RK), Soave, and Peng-Robinson equations. For these equations, u and w take on the integer values listed in Table 5.4. The two critical point conditions may be used to set the values of the two parameters a and b in Eq. (5.55):

$$\left(\frac{\partial p}{\partial V}\right)_{T_c} = 0 \qquad (5.59)$$

and

$$\left(\frac{\partial^2 p}{\partial V^2}\right)_{T_c} = 0 . \qquad (5.60)$$

Eqs. (5.59) and (5.60) are applicable only to pure components. Eqs. (5.59) and (5.60) were used by Soave (1972) and by Peng and Robinson (1976) to find a and b at the critical point. They then made the parameter a a function of the temperature and the acentric factor. The expressions for a and b which result from this procedure are listed in Table 5.4.

Table 5.4 Common Cubic Equations of State with the Extended Corresponding-States Modification $f(\omega,\theta)$

Equation	u	w	B	a
Soave	1	0	$\dfrac{0.08664RT_c}{p_c}$	$\dfrac{0.42748R^2T_c^2}{p_c}\left[1 + f(\omega,\theta)\left(1 - T_r^{1/2}\right)\right]^2$
				where
				$f(\omega,\theta) = 0.48 + 1.574\omega - 27.5\theta$
Peng-Robinson	2	-1	$\dfrac{0.07780RT_c}{p_c}$	$\dfrac{0.45724R^2T_c^2}{p_c}\left[1 + f(\omega,\theta)\left(1 - T_r^{1/2}\right)\right]^2$
				where
				$f(\omega,\theta) = 0.37464 + 1.54226\omega - 42.2\theta$

5.4.2 Corresponding States of the Patel-Teja Equation of State

$$p = \frac{RT}{V - b} - \frac{a(T)}{V(V + b) + c(V - b)} , \qquad (5.61)$$

where

$$a = \Omega_a \frac{R^2 T_c^2}{p_c}\alpha(T) \quad , \quad b = \Omega_b \frac{RT_c}{p_c} , \quad c = \Omega_c \frac{RT_c}{p_c} , \qquad (5.62)$$

$$\left(\frac{\partial p}{\partial V}\right)_{T_c} = 0 \quad , \quad \left(\frac{\partial^2 p}{\partial V^2}\right)_{T_c} = 0 , \quad \frac{p_c V_c}{RT_c} = Z_c , \qquad (5.63)$$

$$\Omega_c = 1 - 3Z_c, \tag{5.64}$$

$$\Omega_a = 3Z_c^2 + 3(1 - 2Z_c)\Omega_b + \Omega_b^2 + \Omega_c, \tag{5.65}$$

$$\Omega_b^3 + (2 - 3Z_c)\Omega_b^2 + 3Z_c^2\Omega_b - Z_c^3 = 0, \tag{5.66}$$

$$f(\omega) = 0.45241 + 1.3098\omega - 46.240\theta, \tag{5.67}$$

$$\xi_c = 0.32903 - 0.076799\omega + 3.3116\theta. \tag{5.68}$$

5.4.3 Departure Functions for Generalized Cubic Equation of State

The Helmholtz free energy and entropy departure functions in Eq (5.55) are as follows:

$$A - A^0 = \frac{a}{b\sqrt{u^2 - 4w}} \ln \frac{2Z + B^*\left(u - \sqrt{u^2 - 4w}\right)}{2Z + B^*\left(u + \sqrt{u^2 - 4w}\right)} - RT \ln \frac{Z - B^*}{Z} - RT \ln \frac{V}{V^0} \tag{5.69}$$

and

$$S - S^0 = R \ln \frac{Z - B^*}{Z} + R \ln \frac{V}{V^0} - \frac{1}{b\sqrt{u^2 - 4w}} \frac{\partial a}{\partial T} \ln \frac{2Z + B^*\left(u - \sqrt{u^2 - 4w}\right)}{2Z + B^*\left(u + \sqrt{u^2 - 4w}\right)}. \tag{5.70}$$

5.5 Hard-Sphere Equation of State

5.5.1 Basic theory

Barker and Henderson (1976) reviewed the scaled-particle theory for hard spheres. For hard spheres of diameter d, the equation of state is given by:

$$pV / Nk_B T = 1 + 4\eta g(d), \tag{5.71}$$

where $\eta = \pi\rho d^3 / 6$, N is the Avagadro constant and k_B the Boltzmann constant.

The factor $\pi d^3 / 6$ is the volume of a hard sphere. Thus, for hard spheres, it is only necessary to find $g(d)$. Reiss et al. (1959) have developed a simple but accurate method for obtaining $g(d)$ and p.

Let $p_0(r)$ be the probability that there is no molecule whose center lies within a sphere of radius r, centered about some specified point. Thus, $-dp_0/dr$ is the probability of finding an empty sphere whose radius lies between r and $r+dr$. This is equal to the product of the probability of having no molecule within the radius r and the conditional probability, $4\pi\rho G(r)r^2 dr$. The significance of $G(r)$ arises from the fact that an empty sphere of radius d affects the remainder of the fluid precisely like another molecule, i.e., $G(d) = g(d)$.

Thus

$$- dp_0(r)/dr = p_0(r)4\pi\rho r^2 G(r) \tag{5.72}$$

Also

$$p_0(r) = \exp\{-\beta W\}, \tag{5.73}$$

where $W(r)$ is the reversible work necessary to create a cavity of radius r in the fluid. It should be noted that p_0, W, and G depend upon ρ as well as r. However, for notational simplicity we do not show this dependence explicitly. Combining Eq. (5.72) and Eq. (5.73) gives:

$$dp_0 / p_0 = -\beta dW = 4\pi\rho G(r)r^2 dr . \tag{5.74}$$

Hence,

$$dW = pdV + \sigma dS = k_B T\rho G(r)dV , \tag{5.75}$$

where σ is the surface tension and S and V are the surface area and the volume of the system, respectively. Therefore,

$$G(r) = (\rho k_B T)^{-1}\left(p + \frac{2\sigma}{r}\right). \tag{5.76}$$

It should be noted that Eq. (5.76) gives:

$$G(\infty) = pV / Nk_B T = 1 + 4\eta G(d). \tag{5.77}$$

To proceed further we need to know the r dependence of σ. For r that is not too small, it is reasonable to assume:

$$\sigma(r) = \sigma_0 \{1 + \delta(d/r)\} \qquad (5.78)$$

where σ_0 and δ are constants to be determined. Substitution of Eq. (5.76) into Eq. (5.74) gives:

$$G(r) = (\rho k_B T)^{-1} \left(p + \frac{2\sigma_0}{r} + \frac{4\sigma_0 \delta d}{r^2} \right). \qquad (5.79)$$

To determine σ_0 and δ, it is necessary to consider $p_0(r)$ again. For $r < d/2$, no more than one molecular center can lie within a sphere of radius r, and therefore, $p_0(r)$ is equal to unity minus the probability of there being a molecular center within the sphere. Thus,

$$p_0(r) = 1 - (4\pi/3)r^3 \rho, \qquad r < d/2. \qquad (5.80)$$

Hence, we have:

$$-dp_0(r)/dr = p_0(r)4\pi r^2 \rho G(r) = 4\pi r^2 \rho \qquad (5.81)$$

for $r < d/2$.
Thus $p_0(r)G(r) = 1$ and

$$G(r) = \left[1 - (4\pi/3)r^3 \rho\right]^{-1}, \qquad r < d/2. \qquad (5.82)$$

Consequently, Eq. (5.82) is valid for all $r < d/2$ and Eq. (5.79) is valid when r is not too small. For $d/2 < r < d/\sqrt{3}$, two molecular centers can lie within the sphere, and $g(r)$ and $p_0(r)$ and $G(r)$ cannot be determined without first knowing $G(r)$.

However, as an approximation it can be assumed that Eq. (5.79) is valid for all $r > d/2$. Hence, approximately:

$$G(r) = \frac{p}{\rho k_B T} + A\left(\frac{d}{r}\right) + B\left(\frac{d}{r}\right)^2, \qquad r < d/2, \qquad (5.83)$$

where $A = 2\sigma_0/\rho K_B Td$ and $B = 4\sigma_0\delta/\rho K_B Td^2$. Solving Eqs. (5.77) and (5.83) for $p/\rho kT$ and $G(r)$ gives:

$$\frac{p}{\rho kT} = \frac{1 + 4\eta(A + B)}{1 - 4\eta} \qquad (5.84)$$

$$G(r) = \left[1 - 8\eta(r/d)^3\right]^{-1}, r < d/2$$

and

$$G(r) = \frac{1 + 4\eta(A+B)}{1 - 4\eta} + \frac{A}{r} + \frac{B}{r^2}, r > d/2 \qquad (5.85)$$

Expressions for A and B, and thus γ_0 and δ can be obtained by requiring that $G(r)$ and dG/dr be continuous at $r = d/2$. The resulting values for A and B are:

$$A = -\frac{3\eta(1+\eta)}{2(1-\eta)^3} \qquad (5.86)$$

and

$$B = \frac{3\eta^2}{4(1-\eta)^3}.$$

Therefore,

$$\frac{p}{\rho k_B T} = \frac{1 + \eta + \eta^2}{(1-\eta)^3}.$$

Thus,

$$g(d) = \frac{4 - 2\eta + \eta^2}{4(1-\eta)^3}. \qquad (5.89)$$

Eq. (5.88) gives an exact second and third virial coefficient and gives reasonable values for the higher virial coefficients also. On the other hand, Eq. (5.88) would inevitably fail at high densities. However, for the densities at which hard spheres are fluid, Eq. (5.88) is in good agreement with the computer simulation results.

5.5.2 Hard Sphere Equation of State

Ree and Hoover (1964) proposed the following Pade approximation from known virial coefficients:

$$Z^{HS} = \frac{p}{\rho k_B T} = \frac{1 + 1.75399\eta + 2.31704\eta^2 + 1.108928\eta^3}{1 - 2.246004\eta + 1.301056\eta^2}. \qquad (5.90)$$

A complete virial expansion of the hard shere equation of state is written as:

$$Z^{HS} = \frac{p}{\rho k_B T} \quad = 1 + 4\eta + 10\eta^2 + 18.36\eta^3 + 28.2\eta^4 + 39.5\eta^5 + ... \quad (5.91)$$

Carnahan and Starling (1969) assumed all coefficients in Eq. (5.91) are integers, and the equation was simplified:

$$Z^{HS} = \frac{p}{\rho k_B T} \quad = 1 + 4\eta + 10\eta^2 + 18\eta^3 + 28\eta^4 + 40\eta^5 + ... \quad (5.92)$$

The coefficients are expressed as:

$$B_n = n^2 + n - 2 \quad (n \geq 2),$$

in which B_n is the nth coefficient. Based on the expression for B_n, Eq. (5.92) can be obtained as:

$$Z^{HS} = \frac{p}{\rho k_B T} \quad = 1 + \sum_{n=2}^{\infty} B_n \eta^{n-1} = 1 + \sum_{n=2}^{\infty} (n^2 + n - 2)\eta^{n-1} = 1 + \sum_{n=0}^{\infty} (n^2 + 3n)\eta^n . \quad (5.93)$$

Some typical hard sphere equations of state are given in Table 5.5.

Table 5.5 Representative Hard Sphere Equations of State

Equation	Compressibility factor, Z^{HS}
Van der Waals	$1/(1-4\eta)$
Percus-Yevick-Thiele	$(1 + 2\eta + 3\eta^2)/(1-\eta)^2 = (1 + \eta + \eta^2 - 3\eta^3)/(1-\eta)^3$
Percus-Yevick-Frisch	$(1 + \eta + \eta^2)/(1-\eta)^3 = (1-\eta)^3/(1-\eta)^4$
Carnahan-Starling	$(1 + \eta + \eta^2 - \eta^3)/(1-\eta)^3$
Kolafa	$(1 + \eta + \eta^2 - 2/3\eta^3 - 2/3\eta^4)/(1-\eta)^3$

Carnahan and Starling (1969) have empirically observed that results even better than those of Eq. (5.88) can be obtained by a simple empirical modification of Eq. (5.88). Carnahan and Starling added the factor $-\eta^3/(1-\eta)^3$ to Eq. (5.88), when the Carnahan-Starling equation of state is obtained:

$$\frac{p}{\rho k_B T} = \frac{1 + \eta + \eta^2 - \eta^3}{(1-\eta)^3}. \quad (5.94)$$

Thus, in this approximation,

$$g(d) = \frac{4 - 2\eta}{4(1 - \eta)^3} .$$ (5.95)

The virial coefficients and equation of state calculated from Eq. (5.94) are in excellent agreement with the exact results. The Carnahan-Starling (CS) expression for the free energy of a hard-sphere fluid may be obtained by integration:

$$\frac{A}{Nk_B T} = 3 \ln \lambda - 1 + \ln \rho + \frac{4\eta - 3\eta^2}{(1 - \eta)^2},$$ (5.96)

where $\lambda = h / (2\pi m k T)^{1/2}$.

5.5.3 Corresponding States of the CS-PT Equation of State

Incorporating Eq. (5.94), the Patel-Teja equation of state is

$$p = \frac{RT}{V} \frac{1 + \eta + \eta^2 - \eta^3}{(1 - \eta)^3} - \frac{a}{V(V + b) + c(V - b)},$$ (5.97)

$$a = \Omega_a \frac{(RT_c)^2}{p_c} \qquad \Omega_a = \Omega_{ac} \alpha(T_r),$$ (5.98)

$$b = \Omega_b \frac{RT_c}{p_c},$$ (5.99)

$$c = \Omega_c \frac{RT_c}{p_c},$$ (5.100)

$$\alpha(T_r) = [1 + f(1 - T_r^{1/2})]^2,$$ (5.101)

and

$$\eta_c = \frac{b}{4V_c},$$ (5.102)

where,

$$\beta_c = \frac{(1 - 4\eta_c - 20\eta_c^2 - 16\eta_c^3 + \eta_c^4) - 2\zeta_c(1 + 2\eta_c)(1 - \eta_c)^4}{4[\zeta_c(1 - \eta_c)^4 - 4\eta_c^5 + 16\eta_c^4 - 18\eta_c^3 - 12\eta_c^2]}$$

$$\gamma_c = \frac{(1 + 4\eta_c + 4\eta_c^2 - 4\eta_c^3 + \eta_c^4)(1 + 4\eta_c + 4\beta_c - 16\eta_c\beta_c)^2}{2(1-\eta_c)^4(1 + 2\eta_c + 2\beta_c)},$$

$$\Omega_{ac} = \gamma_c\zeta_c \qquad \Omega_b = 4\eta_c\zeta_c \qquad \Omega_c = 4\beta_c\zeta_c,$$

$$\zeta_c = 0.3606 + 0.011529\omega - 11.24\theta, \tag{5.103}$$

and

$$f = 0.48265 + 1.4819\omega - 77.7\theta. \tag{5.104}$$

5.6 Martin-Hou Equation of State

5.6.1 Thermodynamic Behavior of the Martin-Hou Equation

Properties of fluids are mainly determined by short-range repulsion of molecules (Henderson, 1979). In the square-well non-spherical hard-particle perturbation theory, the short-range repulsion is expressed by two parts, of which the first is expressed by the hard-particle reference fluid that mainly reflects hard repulsion of molecules. Especially at high densities, the equation of state of the system levels off to that of a hard-particle fluid. The second is expressed by a square-well with a potential energy larger than 0 in the perturbation potential, which mainly reflects soft repulsion of molecules. In these two parts, the former is dominant. To exactly express the hard repulsion of molecules, the first five terms in the hard-particle equation of state are sufficient (Zhang and Hou, 1987).

In the perturbation theory, the Martin-Hou equation of state properly reflects both the perturbation and reference parts: real calculation showed that the Martin-Hou equation of state can well represent not only high-density gas (Hsu and Mcketta, 1964; Smith et al., 1963; Burnside, 1971), but also P-V-T behavior of liquids. The applicable systems include not only hydrocarbon gases but also non-hydrocarbon gases such as water, ammonia and other strong polar substances (Hou et al., 1981). Coefficients of different terms in the Martin-Hou equation of state may be determined by the following thermodynamic behavior:

When $p \to 0$,

$$pV = RT \tag{5.105}$$

at the critical point,

$$(dp/dV)_T = 0 \tag{5.106}$$

$$(d^2 p/dV^2)_T = 0 \tag{5.107}$$

$$(d^3 p/dV^3)_T \le 0$$

$$(d^4 p/dV^4)_T = 0 \tag{5.109}$$

when $T' \cong 0.8T_c$,

$$[(dZ/dp_r)_{T_r}]_{p_r=0} = -(1 - Z_c), \tag{5.110}$$

when the Boyle temperature T_B, $[(dZ/dp_r)_{T_r}]_{p_r=0} = 0$, $\tag{5.111}$

when $V = V_c$,

$$(d^2 p/dT^2)_V = 0, \tag{5.112}$$

when $V = V_c$,

$$(dp/dT)_V = -\alpha_c p_c/T_c, \tag{5.113}$$

where α_c is the Riedel parameter.

Then the Martin-Hou Equation of State is expressed as:

$$p = \frac{RT}{V-b} + \frac{A_2 + B_2 T + C_2 e^{-5.475T/T_c}}{(V-b)^2} + \frac{A_3 + B_3 T + C_3 e^{-5.475T/T_c}}{(V-b)^3}$$
$$+ \frac{A_4}{(V-b)^4} + \frac{B_5 T}{(V-b)^5} \tag{5.114}$$

$$Z_c = \frac{p_c V_c}{RT_c}, \quad b = V_c - \frac{\beta V_c}{15Z_c} \tag{5.115}$$

$$f_2(T_c) = 9p_c(V_c - b)^2 - 3.8RT_c(V_c - b) \tag{5.116}$$

$$f_3(T_c) = 5.4RT_c(V_c - b)^2 - 17p_c(V_c - b)^3 \tag{5.117}$$

$$f_4(T_c) = 12p_c(V_c - b)^4 - 3.4RT_c(V_c - b)^3 \tag{5.118}$$

$$f_s(T_c) = 0.8RT_c(V_c - b)^4 - 3p_c(V_c - b)^5 \tag{5.119}$$

$$C_2 = \frac{\left[f_2(T_c) + bRT' + \dfrac{(RT')^2}{P_c}(1 - Z_c)\right](T_B - T_c) + [f_2(T_c) + bRT_B](T_c - T')}{(T_B - T_c)\left(e^{-5.475} - e^{-5.475T'/T_c}\right) - (T_c - T')\left(e^{-5.475T_B/T_c} - e^{-5.475}\right)} \tag{5.120}$$

$$B_2 = \frac{-f_2(T_c) - bRT_B - C_2\left(e^{-5.475T_B/T_c} - e^{-5.475}\right)}{T_B - T_c} \tag{5.121}$$

$$C_3 = -(V_c - b)C_2 \tag{5.122}$$

$$A_2 = f_2(T_c) - B_2T_c - C_2e^{-5.475} \tag{5.123}$$

$$A_4 = f_4(T_c) \tag{5.124}$$

$$B_5 = \frac{f_5(T_c)}{T_c} \tag{5.125}$$

$$B_3 = -\alpha_c p_c / T_c\left[(V_c - b)^3 - R(V_c - b)^2 - B_2(V_c - b) - B_5/(V_c - b)^2\right] \tag{5.126}$$

$$A_3 = f_3(T_c) - B_3T_c - C_3e^{-5.475}. \tag{5.127}$$

5.6.2 Corresponding States of the Martin-Hou Equation

The Riedel parameter α_c may be obtained from the following correlation (Xiang, 2001a, b, 2002, 2003):

$$\alpha_c = 5.790206 + 4.888195\omega + 33.91196\theta \tag{5.128}$$

The other substance-dependent parameters were determined from Xiang (2003) as follows:

$$T_B / T_c = 2.71 - 1.48\omega - 25.4\theta, \tag{5.129}$$

$$T' / T_c = 0.772 - 0.132\omega - 11.8\theta, \tag{5.130}$$

and

$$\beta = 3.27 - 0.163\omega - 31.9\theta. \tag{5.131}$$

5.7 Equation for Liquids

5.7.1 Introduction

Watson (1943) first attempted to establish an equation for liquid density. However, he did not obtain a generalized equation that could represent the large difference between different kinds of substances. Riedel (1954) proposed a generalized equation for saturated density, in which the Riedel parameter is used as an input parameter. Reid and Sherwood (1966) predicted saturated liquid densities of nonpolar and polar substances, but not of polar associating liquids, in which the Riedel parameter was replaced by the acentric factor. Based on the critical compression factor, Yen and Woods (1966) proposed an equation for the saturated liquid density of nonpolar and polar compounds with a 2.1% calculation error. By means of the critical compression factor, Rackett (1970) also proposed a saturated-liquid-density equation, which is only applicable to nonpolar and polar substances, and not to polar associating compounds. Gunn and Yamada (1971) proposed two different equations, below and above the reduced temperature $T_r = 0.8$, for the saturated liquid density of substances. Lu et al. (1973) employed the Guggenheim (1945) equation to correlate the saturated liquid density of 23 substances with the acentric factor as an input parameter. Spencer and Danner (1972) suggested that a slightly corrected critical compression factor Z_{RA} replace Z_c to correlate the saturated liquid density. Spencer and Adler (1978) further extended the above method to nonpolar and polar fluids; this method, however, gave a relatively large deviation for associating substances. Srinivasan (1989) proposed a two-parameter saturated-liquid-density equation with a relation of the product of the saturated reduced temperature and saturated reduced density to the temperature for low-temperature liquids and refrigerants. Thomson et al. (1982) proposed a method to predict the compressed liquid density by means of a generalization of the Tait equation. Bhirud (1978) proposed a method for polar compounds; however measured density data were required. Hankenson and Thomson (1979) reviewed the above two methods and the method of Lyckman et al. (1965), which was applicable to compressed liquids. However, these methods were more complex than the Hankinson-Brobst-Thomson method and their accuracy seemed to be relatively poor (Thomson et al., 1982). Campbell and Thodos (1985) proposed a method to predict the saturated liquid densities for nonpolar and polar substances, in which a linear relation of Z_{RA} to the reduced temperature was used, and which required the critical parameters and normal boiling point for nonpolar substances, and also the dipole moment for polar substances. For pure substances, the accuracy of this method was similar to that of the Hankinson method and of the modified Rackett method.

5.7.2 Saturated Liquid Density from Corresponding States

According to the critical power law, the saturated-liquid density equation is developed as:

$$\ln \rho_r = d_1 \tau^\beta + d_2 \tau^{\beta+\Delta} + d_3 \tau^{1.5} + d_4 \tau^3 + d_5 \tau^6, \qquad (5.132)$$

where $\tau = 1 - T_r$. $T_r = T/T_c$ is the reduced temperature, with T being the temperature and T_c the critical temperature. Eq. (5.132) contains only three adjustable parameters (d_1, d_2, and d_3) that are dependent on the specific substance and may be determined from the experimental data. $\beta = 0.325$, $\Delta = 0.51$, and $\delta = 4.82$ are the critical exponents derived from the renormalization-group theory of critical phenomena. Eq. (5.132) seems to possess a simple form for describing the liquid density as a function of temperature along the entire vapor-liquid equilibrium curve.

According to the extended corresponding-states principle, the corresponding-states theory describes the saturated liquid density as follows:

$$d_1 = d_{10} + d_{11}\omega + d_{12}\theta$$
$$d_2 = d_{20} + d_{21}\omega + d_{22}\theta$$
$$d_3 = d_{30} + d_{31}\omega + d_{32}\theta$$
$$d_4 = d_{40} + d_{41}\omega + d_{42}\theta$$
$$d_5 = d_{50} + d_{51}\omega + d_{52}\theta.$$

$$(5.133)$$

The general coefficients of Eq. (5.133) are given in Table 5.8.

5.7.3 Comparison with Experimental Data

Figures 5.6 to 5.8 show how the present model represents recent liquid density data for various fluids over their entire temperature range. For most cases the agreement between the model and the experimental data listed in Table 5.8 is generally comparable to the agreement between the different sets of data. For highly accurate data, the present deviations from the experimental data can be almost within 0.1 % and sometimes 0.5%.

As shown in Figs. 5.7 and 5.8, the extended corresponding-states principle can describe the liquid density of refrigerants, hydrogen-bonding and associating substances, such as dichlorodifluoromethane (R12), chlorodifluoromethane (R22), difluoromethane (R32), 1,1-dichloro-2,2,2-trifluoroethane (R123), pentafluoroethane (R125), 1,1,1,2-tetrafluoroethane (R134a), 1,1-difluoroethane (R152a), water, methanol, ammonia, acetone, and acetonitrile. The extended corresponding-states method had an average overall root-mean-square deviation of 1%, and an overall maximum deviation of 5%.

The density of saturated liquid acetonitrile and acetic acid was investigated experimentally by Ter-Gazarian (1906) and by Young (1910) at temperatures up to their critical points, respectively. As discussed by Kratzke and Mueller (1985), the critical temperature given by Ter-Gazarian and by Young seems to be unreliable; therefore, the results for the saturated liquid density are less believable. The critical densities for two substances have been used. In the prediction of the critical density from the new corresponding-states principle, the deviations are 5% and 7% low, which can be thought to be within the acceptable range of experimental uncertainties and to agree reasonably well with the results given by Campbell and Thodos. Therefore, the critical densities for acetonitrile and acetic acid obtained from the extended corresponding-states theory might be reliable. It can be seen from Tables 5.7 and 5.8 that the critical densities for the other substances determined from the extended corresponding-states theory are in agreement with the accurate experimental results. The extended corresponding-states theory provides a reliable method to determine an accurate critical density of a substance.

Table 5.6 Molar Mass M, **Critical Temperature** T_c, **Critical Pressure** p_c, **and Critical Density** ρ_c, **Acentric Factor** ω, **and Aspherical Factor** θ

Substance	M /kg.kmol^{-1}	T_c /K	p_c /kPa	ρ_c /kg.m^{-3}	ω	$\theta \times 10^3$
Argon	39.948	150.69	4863	535	0	0.000
Nitrogen	28.013	126.19	3395	313	0.037	0.000
Methane	16.043	190.564	4598	162	0.011	0.007
Carbon dioxide	44.010	304.13	7377	468	0.225	0.245
Ethylene	28.054	282.35	5041	215	0.086	0.096
Ethane	30.070	305.32	4872	206	0.099	0.097
Propane	44.097	369.83	4248	220	0.152	0.171
Benzene	78.114	562.05	4895	305	0.210	0.472
R12	120.91	385.12	4130	565	0.180	0.196
R22	86.470	369.28	4988	525	0.221	0.501
R32	52.020	351.26	5780	430	0.277	2.560
R123	152.93	456.83	3662	550	0.282	0.481
R125	120.02	339.17	3620	570	0.306	0.388
R134a	102.03	374.18	4055	511	0.327	0.885
R152a	66.050	386.41	4516	369	0.275	1.474

Diethyl ether	74.123	466.7	3640	258	0.281	0.420
Ammonia	17.031	405.4	11345	234	0.256	2.028
Water	18.015	647.1	22050	325	0.344	3.947
Methanol	32.042	512.0	8000	270	0.560	3.486
Acetone	58.080	508.1	4700	270	0.308	2.569
Acetic acid	60.052	593.0	5786	330	0.450	5.844
Acetonitrile	41.053	545.4	4835	225	0.338	9.151

Table 5.7 Data Source of Saturated Liquid Density ρ

Substance	Refs. (ρ)	Refs. (ρ_c)
Argon	Gilgen et al., 1994	535.6 (Gilgen et al., 1994)
Nitrogen	Nowak et al., 1997	313.3 (Nowak et al., 1997; Ambrose and Tsonopoulos, 1995)
Methane	Kleinrahm and Wagner, 1986; McClune, 1976	162.7 ± 0.5 (Kleinrahm and Wagner, 1986)
Carbon dioxide	Duschek et al.,1990	467.6 (Duschek et al., 1990)
Ethylene	Nowak et al., 1996	214.2 (Nowak et al., 1996)
Ethane	McClune, 1976; Haynes and Hiza, 1977	206.6 ± 3 (Ambrose and Tsonopoulos, 1995; Friend et al., 1991)
Propane	McClune, 1976; Haynes and Hiza, 1977; Thomas and Harrison, 1982; Kratzke and Mueller, 1984	220 ± 3 (Ambrose and Tsonopoulos, 1995); 217 (Kudchadker et al., 1968)
Benzene	Goodwin, 1988	305 ± 4 (Goodwin, 1988)
R12	Michels et al., 1966; Haendel et al., 1992	565 (Michels et al., 1966)
R22	Haendel et al., 1992; Kohlen et al., 1985; Fukuizumi and Uematsu, 1992	520 ± 5 (Wagner et al., 1993)
R32	Defibaugh et al., 1994; Holcomb et al., 1993; Higashi, 1994; Magee, 1996	424 ± 5 (Tillner-Roth and Yokozeki, 1997); 427 ± 5 (Higashi, 1994)
R123	Younglove and McLinden, 1994	550 ± 4 (Younglove and McLinden, 1994)
R125	Higashi, 1994; Magee, 1996; Younglove and McLinden, 1994; Defibaugh and Morrison, 1992	572 ± 6 (Higashi, 1994; Piao and Noguchi, 1998); 568 (Sunaga et al., 1998)
R134a	Higashi, 1994; Tillner-Roth	511 ± 3 (Tillner-Roth and

	and Baehr, 1994	Baehr, 1994)
R152a	Holcomb et al., 1993; Blanke and Weiss, 1992; Outcalt and McLinden, 1996; Magee, 1998	368 ± 2 (Outcalt and McLinden, 1996)
Diethyl ether	Kay and Donham, 1955	263 (Young, 1910); 265 (Kay and Donham, 1955)
Ammonia	Kasahara et al., 1999; Streatfeild et al., 1987	234 (Haar and Gallagher, 1978; Cragoe et al., 1922)
Water	IAPWS (1985, 1994)	322 ± 3 (IAPWS, 1985, 1994); 315 ± 15 (Osborne et al., 1939; Kudchadker et al., 1968)
Methanol	Goodwin, 1987; Hales and Ellender, 1976; Machado and Streett, 1983; Yergovich et al., 1971	269 (Goodwin, 1987); 272 (Young, 1910)
Acetone	Yergovich et al., 1971; Campbell and Chatterjee, 1968; Easteal and Woolf, 1982	278 (Kobe et al., 1955); 273 (Kobe and Lynn, 1953)
Acetic acid	Young, 1910	350 (Young, 1910)
Acetonitrile	Malhotra and Woolf, 1991; Ter-Gazarian, 1906; Kratzke and Mueller, 1985	237 (Ter-Gazarian, 1906)

Table 5.8 General Coefficients of Eq. (5.133)

d_{10}	1.416996	d_{11}	0.6407596	d_{12}	44.535700
d_{20}	-0.2505080	d_{21}	-0.9056494	d_{22}	-109.7101
d_{30}	0.03409619	d_{31}	0.6512406	d_{32}	254.2814
d_{40}	0.07416272	d_{41}	0.7161291	d_{42}	-661.1617
d_{50}	0.09119869	d_{51}	-2.706923	d_{52}	1287.925

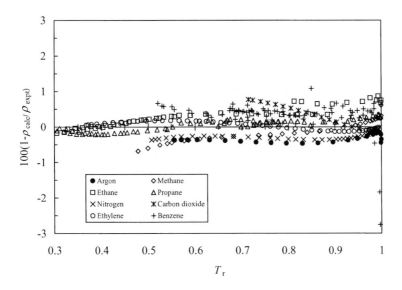

Fig. 5.6 Comparison of the saturated liquid density for nonpolar substances with values calculated from the extended corresponding-states theory, Eq. (5.132)

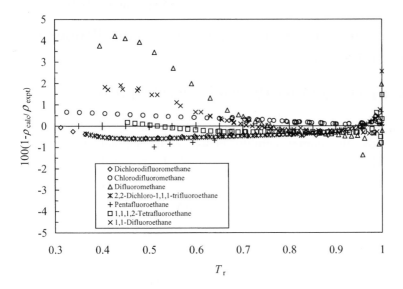

Fig. 5.7 Comparison of the saturated liquid density for chlorofluoro- and hydrofluorocarbons with values calculated from the extended corresponding-states theory, Eq. (5.132)

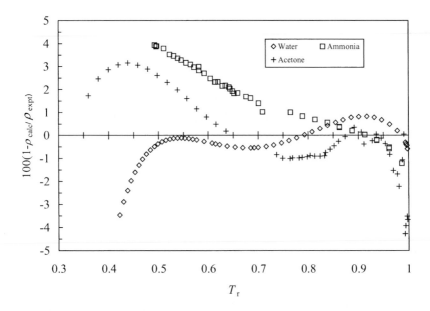

Fig. 5.8 Comparison of the saturated liquid density for some highly polar substances with values calculated from the extended corresponding-states theory, Eq. (5.132)

5.8 Corresponding-States Thermodynamic Properties and Calculated Deviations

5.8.1 Corresponding-States Thermodynamic Properties

According to the basic corresponding-states principle of Van der Waals for spherical fluids, the compressibility factor Z (or another property) may be correlated with the reduced temperature $T_r = T / T_c$, reduced density $\rho_r = \rho / \rho_c$ and reduced pressure $p_r = p / p_c$ with T_c, p_c, and ρ_c the critical parameters to reduce temperature T, pressure p, and density ρ:

$$Z = f_{vdW}(T_r, p_r) = Z^{(0)}. \qquad (5.134)$$

Introduced by Pitzer et al. (1955), a third corresponding-states parameter, the acentric factor, provides a measure of the non-sphericity of a molecule's force field. The following expansion is employed:

$$Z = f_{\text{nonpolar}}(T_r, p_r, \omega) = Z^{(0)} + \omega Z^{(1)}, \qquad (5.135)$$

where $Z^{(0)}$ is for spherical molecules, and the term $Z^{(1)}$ is a deviation function for weakly nonspherical fluids, usually normal fluids. Lee and Kesler (1975) developed a modified BWR equation within the context of the three-parameter correlation of Pitzer et al. The compressibility factor of a real fluid is related to the properties of a simple fluid ($\omega = 0$) and those of n-octane as a reference fluid. The Pitzer aproach to the three-parameter corresponding-states theory assumes that the compressibility factor Z depends linearly on the acentric factor ω. The two points used to establish this straight line are the Z values of a simple fluid and the reference fluid, n-octane. Starling et al. (1972) developed a generalized BWR equation that retained the density dependence given in the original BWR equation. Reid et al. (1987) gave the tabulations of Lee and Kesler (1975). The tabulations of $Z^{(0)}$ and $Z^{(1)}$ were also given by Smith et al. (1996). Teja et al. (1981) used a cubic equation of state.

As given in Chapter 5, the compressibility factor of a highly nonspherical fluid at a reduced temperature and pressure may be represented by the extended corresponding-states principle as follows:

$$Z = f(T_r, p_r, \omega, \theta) = Z^{(0)} + Z^{(1)}\omega + Z^{(2)}\theta, \qquad (5.136)$$

where the $Z^{(2)}$ term is a deviation function for strongly nonspherical fluids.

The procedure for other thermodynamic properties can be similarly obtained according to the extended corresponding-states theory as:

$$\frac{H^0 - H}{RT_c} = \left.\frac{H^0 - H}{RT_c}\right|^{(0)} + \omega\left.\frac{H^0 - H}{RT_c}\right|^{(1)} + \theta\left.\frac{H^0 - H}{RT_c}\right|^{(2)}, \qquad (5.137)$$

$$\frac{S^0 - S}{RT_c} = \left.\frac{S^0 - S}{RT_c}\right|^{(0)} + \omega\left.\frac{S^0 - S}{RT_c}\right|^{(1)} + \theta\left.\frac{S^0 - S}{RT_c}\right|^{(2)}, \qquad (5.138)$$

$$\log(f/p) = \log(f/p)\big|^{(0)} + \omega\log(f/p)\big|^{(1)} + \theta\log(f/p)\big|^{(2)}, \qquad (5.139)$$

and

$$\frac{C_p^0 - C_p}{RT_c} = \left.\frac{C_p^0 - C_p}{RT_c}\right|^{(0)} + \omega\left.\frac{C_p^0 - C_p}{RT_c}\right|^{(1)} + \theta\left.\frac{C_p^0 - C_p}{RT_c}\right|^{(2)}. \qquad (5.140)$$

Comparison results between the calculated and experimental values of the compressibility factor for the vapor and liquid regions are presented for the spherical fluid argon, weakly nonspherical fluids such as methane, nitrogen, ethane, carbon dioxide, chlorodifluoromethane, 1,1,1,2-tetrafluoroethane, and the highly nonspherical polar fluids such as difluoromethane, 1,1,1-trifluoroethane, 1,1-difluoroethane, ammonia, and water.

The spherical, weakly, and highly nonspherical parts of thermodynamic properties may be fitted from the representative substances, which have highly accurate values of their thermodynamic properties from the recent equations of state. The complete thermodynamic properties calculated from the accurate equations of state for these fluids at the selected state points were used to fit $Z^{(0)}$, $Z^{(1)}$, and $Z^{(2)}$, and the other corresponding-states coefficients of the other derived properties at specific points of states are listed in Tables 5.14 to 5.28 (Xiang, 2003). These substances are the spherical fluid argon (Tegeler et al., 1999), weakly nonspherical fluids such as methane (Setzmann and Wagner, 1991), nitrogen (Span et al., 2000), ethane (Friend et al., 1991), carbon dioxide (Span and Wagner, 1996), chlorodifluoromethane (Wagner et al., 1993), and 1,1,1,2-tetrafluoroethane (Tillner-Roth and Baehr, 1994), and highly nonspherical polar fluids, including difluoromethane (Tillner-Roth and Yokozeki, 1997), 1,1,1-trifluoroethane (Lemmon and Jacobsen, 2000), 1,1-difluoroethane (Outcalt and McLinden, 1996), ammonia (Baehr and Tillner-Roth, 1995), and water (Wagner and Pruss, 2002). The corresponding-states coefficients listed in Tables 5.14 to 5.28 may be used not only to interpolate the values for calculation, but also to obtain the deviations of the present theory from experiments, as shown below in Section 5.8.2.

Alternatively, in principle, a spherical fluid, a weakly nonspherical fluid, and a highly nonsphercial fluid may be used to establish for a property of the extended corresponding-states method to continuously calculate the corresponding property at the physical state of a temperature, density or pressure. Consequently, for example, the compressibility factor given in Eq. (5.136) can be obtained as:

$$Z^{(1)} = \frac{(Z_2 - Z_0)\theta_1 - (Z_1 - Z_0)\theta_2}{\omega_2\theta_1 - \omega_1\theta_2} \qquad (5.141)$$

$$Z^{(2)} = \frac{(Z_1 - Z_0)\omega_2 - (Z_2 - Z_0)\omega_1}{\omega_2\theta_1 - \omega_1\theta_2}, \qquad (5.142)$$

where Z_0 is the compressibility factor of the spherical fluid argon, which may be obtained from the equation of state (Tegeler et al., 1999). Z_1, ω_1, and θ_1 are its compressibility factor, acentric factor, and aspherical factor for weakly non-spherical fluids, such as 1, 1, 1, 2-tetrafluoroethane (Tillner-Roth and Baehr, 1994), and Z_2, ω_2, and θ_2 are the compressibility factor, acentric factor, and aspherical factor of strongly polar fluids, such as water (Wagner and Pruss, 2002).

Table 5.14 Values of $Z^{(0)}$

T_r / p_r	0.01	0.05	0.1	0.2	0.4	0.6	0.8	1
0.55	0.9817	0.0099	0.0198	0.0396	0.0791	0.1184	0.1576	0.1967
0.60	0.9857	0.0094	0.0188	0.0375	0.0749	0.1121	0.1491	0.1860
0.65	0.9886	0.0090	0.0180	0.0359	0.0716	0.1070	0.1424	0.1775
0.70	0.9908	0.9521	0.0174	0.0347	0.0691	0.1033	0.1372	0.1710
0.75	0.9924	0.9608	0.9183	0.0338	0.0674	0.1006	0.1335	0.1662
0.80	0.9937	0.9676	0.9331	0.8557	0.0665	0.0992	0.1314	0.1633
0.85	0.9947	0.9729	0.9444	0.8824	0.0669	0.0993	0.1312	0.1626
0.90	0.9955	0.9771	0.9533	0.9024	0.7811	0.1022	0.1339	0.1649
0.93	0.9959	0.9792	0.9577	0.9121	0.8070	0.6639	0.1383	0.1688
0.95	0.9961	0.9804	0.9603	0.9179	0.8216	0.6981	0.1441	0.1734
0.97	0.9964	0.9816	0.9627	0.9231	0.8346	0.7256	0.5604	0.1813
0.98	0.9965	0.9822	0.9639	0.9256	0.8406	0.7376	0.5922	0.1878
0.99	0.9966	0.9827	0.9649	0.9279	0.8462	0.7487	0.6176	0.1989
1.00	0.9967	0.9832	0.9660	0.9302	0.8515	0.7589	0.6391	0.2893
1.01	0.9968	0.9837	0.9670	0.9323	0.8566	0.7685	0.6578	0.4767
1.02	0.9969	0.9842	0.9680	0.9344	0.8614	0.7774	0.6745	0.5247
1.05	0.9971	0.9855	0.9707	0.9401	0.8745	0.8011	0.7157	0.6096
1.10	0.9975	0.9874	0.9745	0.9483	0.8929	0.8327	0.7665	0.6920
1.15	0.9978	0.9890	0.9778	0.9551	0.9077	0.8575	0.8040	0.7466
1.2	0.9981	0.9903	0.9806	0.9608	0.9200	0.8775	0.8331	0.7869
1.3	0.9985	0.9925	0.9850	0.9698	0.9390	0.9076	0.8758	0.8436
1.4	0.9988	0.9941	0.9883	0.9766	0.9530	0.9293	0.9057	0.8823
1.5	0.9991	0.9954	0.9908	0.9816	0.9633	0.9451	0.9272	0.9097
1.6	0.9993	0.9963	0.9927	0.9855	0.9712	0.9571	0.9434	0.9301
1.7	0.9994	0.9971	0.9942	0.9885	0.9773	0.9664	0.9559	0.9457
1.8	0.9995	0.9977	0.9954	0.9910	0.9822	0.9737	0.9656	0.9579
1.9	0.9996	0.9982	0.9964	0.9929	0.9861	0.9796	0.9734	0.9676
2.0	0.9997	0.9986	0.9972	0.9945	0.9893	0.9843	0.9797	0.9754
2.2	0.9998	0.9992	0.9984	0.9969	0.9940	0.9914	0.9890	0.9869
2.4	0.9999	0.9996	0.9993	0.9986	0.9973	0.9963	0.9954	0.9948
2.6	1.0000	0.9999	0.9998	0.9997	0.9995	0.9997	1.0000	1.0000
2.8	1.0000	1.0000	1.0000	1.0006	1.0011	1.0021	1.0029	1.0038
3.0	1.0000	1.0000	1.0006	1.0013	1.0025	1.0037	1.0053	1.0069
3.5	1.0000	1.0006	1.0010	1.0020	1.0042	1.0065	1.0087	1.0110
4.0	1.0000	1.0010	1.0010	1.0026	1.0052	1.0075	1.0101	1.0127

Table 5.14 Values of $Z^{(0)}$ (continued)

T_r / p_r	1.2	1.5	2	3	5	7	10
0.55	0.2356	0.2938	0.3902	0.5810	0.9552	1.3209	1.8569
0.60	0.2227	0.2776	0.3684	0.5476	0.8983	1.2403	1.7392
0.65	0.2125	0.2646	0.3508	0.5207	0.8516	1.1732	1.6413
0.70	0.2046	0.2545	0.3370	0.4989	0.8132	1.1170	1.5586
0.75	0.1986	0.2468	0.3261	0.4814	0.7813	1.0703	1.4886
0.80	0.1949	0.2417	0.3185	0.4682	0.7556	1.0314	1.4287
0.85	0.1935	0.2393	0.3140	0.4587	0.7350	0.9988	1.3779
0.90	0.1953	0.2401	0.3129	0.4530	0.7190	0.9719	1.3346
0.93	0.1987	0.2428	0.3142	0.4516	0.7115	0.9582	1.3115
0.95	0.2026	0.2458	0.3161	0.4514	0.7073	0.9500	1.2975
0.97	0.2086	0.2503	0.3191	0.4521	0.7038	0.9426	1.2843
0.98	0.2129	0.2533	0.3210	0.4526	0.7023	0.9391	1.2775
0.99	0.2184	0.2569	0.3233	0.4534	0.7010	0.9357	1.2713
1.00	0.2262	0.2613	0.3259	0.4545	0.6998	0.9326	1.2652
1.01	0.2378	0.2667	0.3290	0.4557	0.6987	0.9295	1.2597
1.02	0.2581	0.2735	0.3326	0.4572	0.6978	0.9268	1.2539
1.05	0.4562	0.3086	0.3472	0.4632	0.6962	0.9191	1.2385
1.10	0.6065	0.4618	0.3915	0.4792	0.6966	0.9094	1.2156
1.15	0.6850	0.5870	0.4712	0.5045	0.7010	0.9032	1.1962
1.2	0.7389	0.6654	0.5613	0.5399	0.7094	0.9000	1.1806
1.3	0.8114	0.7639	0.6930	0.6309	0.7369	0.9021	1.1570
1.4	0.8593	0.8259	0.7763	0.7186	0.7751	0.9130	1.1421
1.5	0.8927	0.8684	0.8325	0.7872	0.8174	0.9302	1.1352
1.6	0.9173	0.8993	0.8728	0.8391	0.8581	0.9506	1.1331
1.7	0.9360	0.9226	0.9031	0.8786	0.8942	0.9718	1.1340
1.8	0.9506	0.9405	0.9263	0.9093	0.9249	0.9921	1.1376
1.9	0.9621	0.9547	0.9445	0.9334	0.9506	1.0103	1.1422
2.0	0.9714	0.9660	0.9591	0.9527	0.9720	1.0266	1.1473
2.2	0.9850	0.9827	0.9802	0.9809	1.0042	1.0531	1.1563
2.4	0.9943	0.9940	0.9946	1.0000	1.0268	1.0722	1.1636
2.6	1.0007	1.0019	1.0045	1.0128	1.0417	1.0851	1.1687
2.8	1.0054	1.0076	1.0116	1.0223	1.0529	1.0947	1.1722
3.0	1.0088	1.0113	1.0163	1.0288	1.0605	1.1008	1.1733
3.5	1.0136	1.0170	1.0239	1.0377	1.0705	1.1086	1.1724
4.0	1.0156	1.0199	1.0267	1.0416	1.0741	1.1093	1.1671

Table 5.15 Values of $Z^{(1)}$

T_r / p_r	0.01	0.05	0.1	0.2	0.4	0.6	0.8	1
0.55	0.0044	-0.0059	-0.0117	-0.0234	-0.0465	-0.0692	-0.0916	-0.1132
0.60	-0.0314	-0.0054	-0.0109	-0.0215	-0.0425	-0.0632	-0.0835	-0.1036
0.65	-0.0197	-0.0050	-0.0100	-0.0199	-0.0392	-0.0580	-0.0766	-0.0949
0.70	-0.0134	-0.0691	-0.0094	-0.0186	-0.0367	-0.0546	-0.0715	-0.0885
0.75	-0.0074	-0.0375	-0.0795	-0.0168	-0.0325	-0.0476	-0.0626	-0.0768
0.80	-0.0051	-0.0255	-0.0521	-0.1166	-0.0313	-0.0456	-0.0593	-0.0724
0.85	-0.0034	-0.0174	-0.0346	-0.0737	-0.0310	-0.0446	-0.0575	-0.0692
0.90	-0.0026	-0.0116	-0.0232	-0.0473	-0.1081	-0.0456	-0.0567	-0.0670
0.93	-0.0019	-0.0091	-0.0182	-0.0362	-0.0769	-0.1490	-0.0578	-0.0664
0.95	-0.0015	-0.0075	-0.0152	-0.0302	-0.0616	-0.1051	-0.0609	-0.0661
0.97	-0.0012	-0.0063	-0.0125	-0.0249	-0.0489	-0.0766	-0.1404	-0.0665
0.98	-0.0014	-0.0061	-0.0113	-0.0226	-0.0438	-0.0656	-0.1000	-0.0671
0.99	-0.0014	-0.0052	-0.0103	-0.0203	-0.0390	-0.0566	-0.0752	-0.0682
1.00	-0.0011	-0.0046	-0.0094	-0.0182	-0.0344	-0.0484	-0.0585	-0.0693
1.01	-0.0009	-0.0044	-0.0085	-0.0161	-0.0304	-0.0411	-0.0458	-0.0214
1.02	-0.0008	-0.0038	-0.0077	-0.0145	-0.0264	-0.0345	-0.0353	-0.0092
1.05	-0.0003	-0.0029	-0.0053	-0.0098	-0.0168	-0.0190	-0.0135	0.0111
1.10	-0.0004	-0.0015	-0.0024	-0.0043	-0.0050	-0.0011	0.0089	0.0305
1.15	0.0001	-0.0007	-0.0008	-0.0001	0.0032	0.0103	0.0227	0.0421
1.2	-0.0002	0.0004	0.0009	0.0027	0.0088	0.0180	0.0315	0.0499
1.3	0.0002	0.0010	0.0027	0.0062	0.0150	0.0267	0.0405	0.0576
1.4	0.0002	0.0016	0.0030	0.0069	0.0165	0.0281	0.0412	0.0566
1.5	0.0002	0.0018	0.0036	0.0077	0.0175	0.0287	0.0417	0.0556
1.6	0.0003	0.0019	0.0037	0.0079	0.0175	0.0283	0.0407	0.0536
1.7	0.0004	0.0019	0.0039	0.0078	0.0168	0.0271	0.0384	0.0509
1.8	0.0005	0.0017	0.0035	0.0071	0.0164	0.0261	0.0364	0.0474
1.9	0.0003	0.0015	0.0034	0.0071	0.0152	0.0243	0.0340	0.0440
2.0	0.0004	0.0014	0.0032	0.0068	0.0140	0.0225	0.0317	0.0414
2.2	0.0003	0.0014	0.0030	0.0058	0.0123	0.0197	0.0274	0.0359
2.4	0.0003	0.0012	0.0020	0.0043	0.0109	0.0157	0.0227	0.0295
2.6	0.0000	0.0004	0.0030	0.0028	0.0089	0.0128	0.0198	0.0268
2.8	0.0000	0.0000	0.0015	0.0045	0.0092	0.0117	0.0179	0.0238
3.0	0.0000	0.0025	0.0020	0.0018	0.0084	0.0127	0.0169	0.0198
3.5	0.0000	0.0010	0.0025	0.0025	0.0052	0.0084	0.0112	0.0159
4.0	0.0000	-0.0020	0.0025	-0.0005	0.0028	0.0069	0.0101	0.0131

Table 5.15 Values of $Z^{(1)}$ (continued)

$T_r \diagup p_r$	1.2	1.5	2	3	5	7	10
0.55	-0.1351	-0.1671	-0.2190	-0.3170	-0.4972	-0.6559	-0.8730
0.60	-0.1230	-0.1517	-0.1984	-0.2866	-0.4435	-0.5831	-0.7712
0.65	-0.1127	-0.1384	-0.1804	-0.2588	-0.3985	-0.5211	-0.6790
0.70	-0.1046	-0.1284	-0.1664	-0.2356	-0.3586	-0.4635	-0.6013
0.75	-0.0905	-0.1107	-0.1422	-0.2003	-0.3023	-0.3898	-0.5058
0.80	-0.0852	-0.1030	-0.1317	-0.1833	-0.2723	-0.3488	-0.4459
0.85	-0.0807	-0.0970	-0.1218	-0.1664	-0.2433	-0.3081	-0.3915
0.90	-0.0769	-0.0903	-0.1115	-0.1495	-0.2146	-0.2706	-0.3421
0.93	-0.0740	-0.0861	-0.1046	-0.1384	-0.1973	-0.2477	-0.3132
0.95	-0.0725	-0.0827	-0.0991	-0.1304	-0.1853	-0.2330	-0.2943
0.97	-0.0699	-0.0782	-0.0930	-0.1220	-0.1734	-0.2184	-0.2755
0.98	-0.0679	-0.0751	-0.0894	-0.1167	-0.1670	-0.2109	-0.2675
0.99	-0.0649	-0.0714	-0.0851	-0.1122	-0.1608	-0.2032	-0.2575
1.00	-0.0597	-0.0670	-0.0804	-0.1069	-0.1545	-0.1959	-0.2472
1.01	-0.0471	-0.0606	-0.0751	-0.1015	-0.1480	-0.1881	-0.2402
1.02	-0.0021	-0.0509	-0.0689	-0.0953	-0.1413	-0.1808	-0.2280
1.05	0.1032	0.0225	-0.0413	-0.0755	-0.1210	-0.1577	-0.2037
1.10	0.0707	0.1656	0.0528	-0.0306	-0.0837	-0.1195	-0.1635
1.15	0.0711	0.1332	0.1623	0.0303	-0.0428	-0.0801	-0.1194
1.2	0.0738	0.1198	0.1842	0.0986	0.0011	-0.0401	-0.0822
1.3	0.0768	0.1102	0.1666	0.1893	0.0905	0.0385	-0.0042
1.4	0.0732	0.1007	0.1476	0.2061	0.1638	0.1123	0.0689
1.5	0.0707	0.0948	0.1358	0.2007	0.2102	0.1714	0.1299
1.6	0.0674	0.0888	0.1261	0.1897	0.2331	0.2148	0.1794
1.7	0.0633	0.0830	0.1168	0.1782	0.2411	0.2439	0.2236
1.8	0.0589	0.0773	0.1085	0.1665	0.2406	0.2590	0.2535
1.9	0.0551	0.0717	0.1005	0.1559	0.2355	0.2702	0.2772
2.0	0.0512	0.0669	0.0932	0.1455	0.2267	0.2720	0.2915
2.2	0.0442	0.0577	0.0814	0.1297	0.2112	0.2658	0.3123
2.4	0.0397	0.0513	0.0699	0.1142	0.1923	0.2534	0.3161
2.6	0.0339	0.0434	0.0643	0.1035	0.1786	0.2398	0.3112
2.8	0.0305	0.0389	0.0554	0.0909	0.1614	0.2243	0.2990
3.0	0.0238	0.0341	0.0516	0.0824	0.1479	0.2131	0.2910
3.5	0.0185	0.0279	0.0382	0.0668	0.1244	0.1810	0.2626
4.0	0.0145	0.0202	0.0305	0.0541	0.1041	0.1582	0.2360

Table 5.16 Values of $Z^{(2)}$

T_r / p_r	0.01	0.05	0.1	0.2	0.4	0.6	0.8	1
0.55	-20.118	-0.2911	-0.5834	-1.1693	-2.3618	-3.5631	-4.7975	-6.0793
0.60	-1.4400	-0.3006	-0.5772	-1.1823	-2.4081	-3.6420	-4.8879	-6.1548
0.65	-0.6345	-0.3041	-0.6108	-1.2259	-2.4754	-3.7421	-5.0307	-6.3136
0.70	-0.2393	-2.8433	-0.6171	-1.2427	-2.5041	-3.7707	-5.0715	-6.3620
0.75	-0.2410	-1.8189	-5.7648	-1.2891	-2.7165	-4.1486	-5.5637	-7.0234
0.80	-0.1137	-0.8794	-2.5102	-10.1462	-2.7746	-4.1985	-5.6362	-7.0867
0.85	-0.0449	-0.3936	-1.1549	-3.8798	-2.8056	-4.2449	-5.6890	-7.1695
0.90	0.0004	-0.1529	-0.4670	-1.6200	-7.3662	-4.2927	-5.8069	-7.3292
0.93	-0.0049	-0.0436	-0.2092	-0.8873	-4.0182	-13.814	-5.9451	-7.4760
0.95	0.0166	-0.0095	-0.1038	-0.5487	-2.6805	-7.6214	-6.0610	-7.6433
0.97	0.0141	0.0219	-0.0073	-0.2797	-1.7732	-4.8190	-13.515	-7.9196
0.98	0.0200	0.0529	0.0075	-0.1801	-1.3957	-3.8807	-8.8367	-8.1215
0.99	0.0318	0.0481	0.0337	-0.0989	-1.0693	-3.1189	-6.7392	-8.3859
1.00	0.0238	0.0486	0.0982	-0.0266	-0.8116	-2.4866	-5.3195	-10.577
1.01	0.0054	0.0940	0.1058	0.0387	-0.5607	-1.9872	-4.2557	-7.2186
1.02	0.0114	0.0710	0.1426	0.1016	-0.3804	-1.5664	-3.4475	-5.9707
1.05	-0.0059	0.1222	0.1732	0.2338	0.0476	-0.6168	-1.7468	-3.4290
1.10	0.0303	0.1451	0.2255	0.4093	0.4630	0.1942	-0.3595	-1.3271
1.15	0.0275	0.1731	0.2876	0.4804	0.6598	0.5873	0.2682	-0.3689
1.2	0.0513	0.1638	0.2741	0.5089	0.7431	0.7664	0.5321	0.0345
1.3	0.0308	0.1518	0.2679	0.4937	0.7530	0.7879	0.6626	0.2574
1.4	0.0483	0.2004	0.4072	0.7644	1.2570	1.5392	1.6129	1.4739
1.5	0.0524	0.1771	0.3150	0.6055	1.0006	1.2134	1.2052	1.0698
1.6	0.0211	0.1319	0.2641	0.4714	0.7711	0.9041	0.8326	0.6593
1.7	0.0151	0.0912	0.2016	0.3837	0.6042	0.6542	0.5607	0.3023
1.8	-0.0066	0.0795	0.1707	0.3067	0.4140	0.3848	0.2633	0.0298
1.9	0.0170	0.0760	0.1282	0.2073	0.2677	0.2116	0.0535	-0.2214
2.0	0.0151	0.0625	0.0894	0.1455	0.1771	0.0675	-0.1599	-0.4794
2.2	0.0170	0.0177	0.0117	0.0316	-0.0386	-0.2094	-0.4742	-0.8292
2.4	0.0016	-0.0152	-0.0104	-0.0444	-0.0879	-0.2498	-0.6594	-0.8353
2.6	0.0000	-0.0142	-0.2588	0.0255	-0.2188	-0.3691	-0.7813	-1.1636
2.8	0.0000	0.0000	0.0761	-0.3152	-0.3631	-0.5998	-0.8800	-1.3542
3.0	0.0000	-0.2366	-0.0786	0.0349	-0.6120	-0.8137	-1.1141	-1.5054
3.5	0.0000	-0.2140	-0.2366	-0.2366	-0.5144	-0.9238	-1.2016	-1.6567
4.0	0.0000	0.1773	-0.2366	-0.1538	-0.5896	-0.9999	-1.4357	-1.8270

Table 5.16 Values of $Z^{(2)}$ (continued)

$T_r p_r$	1.2	1.5	2	3	5	7	10
0.55	-7.3247	-9.2700	-12.563	-19.506	-34.239	-50.208	-75.197
0.60	-7.4586	-9.4040	-12.741	-19.563	-34.631	-50.399	-74.597
0.65	-7.6448	-9.6548	-13.044	-20.034	-34.697	-50.234	-74.638
0.70	-7.6843	-9.6712	-13.033	-20.239	-34.950	-50.294	-73.866
0.75	-8.4747	-10.656	-14.319	-21.843	-37.003	-52.729	-76.175
0.80	-8.5336	-10.758	-14.435	-21.896	-36.943	-52.305	-75.408
0.85	-8.5949	-10.824	-14.587	-22.112	-37.095	-52.202	-74.793
0.90	-8.8213	-11.097	-14.871	-22.326	-37.264	-52.005	-74.049
0.93	-9.0533	-11.321	-15.116	-22.607	-37.402	-52.046	-73.756
0.95	-9.2050	-11.502	-15.317	-22.790	-37.501	-52.036	-73.625
0.97	-9.4843	-11.761	-15.564	-22.988	-37.616	-52.001	-73.497
0.98	-9.6701	-11.946	-15.695	-23.165	-37.693	-52.013	-73.111
0.99	-9.8891	-12.131	-15.854	-23.250	-37.776	-52.034	-73.053
1.00	-10.1663	-12.314	-16.028	-23.404	-37.833	-52.027	-73.068
1.01	-10.5497	-12.548	-16.193	-23.507	-37.880	-52.026	-72.912
1.02	-10.2124	-12.819	-16.383	-23.677	-37.973	-52.042	-73.106
1.05	-5.9826	-12.135	-16.872	-24.050	-38.151	-52.066	-72.676
1.10	-2.8246	-6.7924	-15.325	-24.412	-38.411	-51.972	-71.887
1.15	-1.4114	-3.9758	-10.849	-23.615	-38.361	-51.768	-71.448
1.2	-0.7409	-2.6345	-7.8991	-21.140	-37.861	-51.331	-70.508
1.3	-0.3295	-1.6740	-5.0999	-15.390	-35.180	-49.476	-68.617
1.4	1.1519	0.2270	-2.2932	-10.522	-30.912	-47.065	-67.177
1.5	0.7189	-0.0948	-2.2106	-8.8584	-26.810	-43.323	-63.730
1.6	0.2944	-0.4385	-2.3841	-8.0928	-23.868	-39.786	-60.057
1.7	-0.0456	-0.7938	-2.5952	-7.7648	-21.851	-36.902	-56.923
1.8	-0.3324	-1.1105	-2.8249	-7.5732	-20.478	-34.440	-54.063
1.9	-0.6436	-1.3571	-3.0375	-7.5157	-19.503	-32.750	-51.612
2.0	-0.8456	-1.6202	-3.2139	-7.4937	-18.673	-31.249	-49.293
2.2	-1.2171	-1.9600	-3.5468	-7.6130	-17.689	-29.210	-46.138
2.4	-1.6073	-2.2854	-3.7647	-7.7364	-17.099	-27.536	-43.588
2.6	-1.6848	-2.2934	-4.0119	-7.7033	-16.634	-26.253	-41.551
2.8	-2.0274	-2.5985	-4.0018	-7.4521	-15.856	-25.199	-39.455
3.0	-1.8022	-2.5915	-4.3074	-7.6000	-15.545	-24.699	-38.194
3.5	-2.0477	-2.9578	-4.2940	-7.4055	-14.699	-22.732	-35.312
4.0	-2.199	-2.9667	-4.3028	-7.2530	-14.049	-21.340	-32.805

Table 5.17 Values of $(H^0 - H)/RT_c \big|^{(0)}$

T_r / p_r	0.01	0.05	0.1	0.2	0.4	0.6	0.8	1
0.55	0.0258	5.2740	5.2708	5.2637	5.2489	5.2347	5.2203	5.2056
0.60	0.0216	5.1347	5.1315	5.1245	5.1111	5.0975	5.0844	5.0700
0.65	0.0195	4.9901	4.9871	4.9811	4.9688	4.9566	4.9444	4.9318
0.70	0.0174	0.0913	4.8346	4.8296	4.8194	4.8092	4.7986	4.7874
0.75	0.0157	0.0816	0.1713	4.6619	4.6550	4.6476	4.6396	4.6311
0.80	0.0141	0.0724	0.1504	0.3293	4.4774	4.4752	4.4719	4.4675
0.85	0.0127	0.0648	0.1336	0.2861	4.2688	4.2763	4.2807	4.2830
0.90	0.0115	0.0586	0.1200	0.2528	0.5798	4.0238	4.0477	4.0648
0.93	0.0109	0.0553	0.1130	0.2363	0.5291	0.9500	3.8659	3.9040
0.95	0.0105	0.0533	0.1086	0.2264	0.5006	0.8698	3.7049	3.7734
0.97	0.0101	0.0514	0.1047	0.2173	0.4753	0.8068	1.3499	3.6060
0.98	0.0100	0.0505	0.1027	0.2129	0.4636	0.7797	1.2590	3.4960
0.99	0.0098	0.0496	0.1009	0.2087	0.4526	0.7550	1.1885	3.3428
1.00	0.0096	0.0488	0.0991	0.2047	0.4421	0.7322	1.1307	2.7039
1.01	0.0095	0.0479	0.0973	0.2009	0.4321	0.7110	1.0808	1.7619
1.02	0.0093	0.0471	0.0956	0.1971	0.4225	0.6913	1.0381	1.5932
1.05	0.0089	0.0449	0.0909	0.1866	0.3964	0.6392	0.9339	1.3269
1.10	0.0082	0.0414	0.0838	0.1713	0.3593	0.5693	0.8096	1.0940
1.15	0.0076	0.0384	0.0776	0.1579	0.3283	0.5139	0.7187	0.9481
1.20	0.0071	0.0358	0.0721	0.1463	0.3019	0.4684	0.6479	0.8423
1.30	0.0062	0.0312	0.0628	0.1269	0.2592	0.3973	0.5417	0.6926
1.40	0.0054	0.0273	0.0548	0.1104	0.2241	0.3410	0.4612	0.5845
1.50	0.0048	0.0243	0.0486	0.0977	0.1973	0.2987	0.4017	0.5061
1.60	0.0043	0.0217	0.0434	0.0871	0.1752	0.2643	0.3539	0.4442
1.70	0.0039	0.0195	0.0390	0.0781	0.1567	0.2355	0.3147	0.3937
1.80	0.0035	0.0176	0.0352	0.0703	0.1408	0.2113	0.2816	0.3516
1.90	0.0032	0.0159	0.0318	0.0636	0.1271	0.1903	0.2533	0.3158
2.00	0.0029	0.0144	0.0289	0.0577	0.1151	0.1721	0.2288	0.2849
2.2	0.0024	0.0120	0.0239	0.0478	0.0951	0.1420	0.1883	0.2341
2.4	0.0020	0.0100	0.0200	0.0398	0.0792	0.1180	0.1563	0.1940
2.6	0.0017	0.0084	0.0167	0.0333	0.0661	0.0985	0.1303	0.1615
2.8	0.0014	0.0070	0.0140	0.0279	0.0553	0.0822	0.1087	0.1346
3.0	0.0012	0.0059	0.0117	0.0233	0.0461	0.0686	0.0905	0.1120
3.5	0.0007	0.0037	0.0073	0.0146	0.0288	0.0426	0.0561	0.0692
4.0	0.0004	0.0021	0.0042	0.0083	0.0163	0.0240	0.0314	0.0385

Table 5.17 Values of $(H^0 - H)/RT_c\big|^{(0)}$ **(continued)**

T_r / p_r	1.2	1.5	2	3	5	7	10
0.55	5.1911	5.1690	5.1326	5.0577	4.9068	4.7534	4.5198
0.60	5.0563	5.0359	5.0004	4.9291	4.7830	4.6330	4.4033
0.65	4.9194	4.8998	4.8671	4.7999	4.6596	4.5146	4.2896
0.70	4.7762	4.7592	4.7299	4.6684	4.5364	4.3960	4.1775
0.75	4.6229	4.6091	4.5842	4.5301	4.4084	4.2755	4.0639
0.80	4.4622	4.4531	4.4352	4.3909	4.2829	4.1587	3.9550
0.85	4.2837	4.2825	4.2749	4.2452	4.1541	4.0400	3.8460
0.90	4.0777	4.0896	4.0990	4.0903	4.0225	3.9204	3.7370
0.93	3.9300	3.9571	3.9826	3.9925	3.9417	3.8481	3.6717
0.95	3.8164	3.8594	3.8992	3.9244	3.8866	3.7992	3.6276
0.97	3.6832	3.7500	3.8104	3.8541	3.8309	3.7503	3.5841
0.98	3.6049	3.6899	3.7632	3.8176	3.8024	3.7259	3.5631
0.99	3.5149	3.6254	3.7144	3.7804	3.7740	3.7008	3.5412
1.00	3.4079	3.5551	3.6633	3.7424	3.7457	3.6763	3.5191
1.01	3.2728	3.4779	3.6095	3.7043	3.7167	3.6517	3.4973
1.02	3.0804	3.3920	3.5531	3.6649	3.6876	3.6268	3.4756
1.05	1.9801	3.0511	3.3656	3.5416	3.5995	3.5513	3.4104
1.10	1.4467	2.1656	2.9652	3.3161	3.4478	3.4251	3.3015
1.15	1.2088	1.6727	2.4699	3.0628	3.2893	3.2963	3.1925
1.2	1.0543	1.4073	2.0435	2.7891	3.1265	3.1663	3.0842
1.3	0.8503	1.0988	1.5304	2.2535	2.7906	2.9037	2.8686
1.4	0.7108	0.9049	1.2336	1.8316	2.4573	2.6403	2.6553
1.5	0.6118	0.7718	1.0382	1.5326	2.1542	2.3879	2.4476
1.6	0.5348	0.6708	0.8949	1.3115	1.8915	2.1525	2.2480
1.7	0.4728	0.5906	0.7833	1.1412	1.6683	1.9376	2.0598
1.8	0.4213	0.5248	0.6931	1.0055	1.4801	1.7451	1.8836
1.9	0.3778	0.4696	0.6182	0.8935	1.3203	1.5734	1.7200
2.0	0.3404	0.4223	0.5547	0.7993	1.1837	1.4207	1.5687
2.2	0.2792	0.3455	0.4522	0.6485	0.9611	1.1633	1.3014
2.4	0.2311	0.2855	0.3726	0.5324	0.7876	0.9567	1.0747
2.6	0.1922	0.2371	0.3087	0.4396	0.6481	0.7867	0.8825
2.8	0.1600	0.1972	0.2562	0.3636	0.5332	0.6447	0.7176
3.0	0.1330	0.1637	0.2122	0.3000	0.4367	0.5242	0.5749
3.5	0.0819	0.1002	0.1288	0.1790	0.2512	0.2886	0.2877
4.0	0.0453	0.0549	0.0695	0.0931	0.1191	0.1192	0.0768

Table 5.18 Values of $(H^0 - H)/RT_c \big|^{(1)}$

T_r / p_r	0.01	0.05	0.1	0.2	0.4	0.6	0.8	1
0.55	-0.1652	8.4401	8.4470	8.4514	8.4692	8.4829	8.4954	8.5112
0.60	0.1253	7.7746	7.7783	7.7854	7.7984	7.8105	7.8234	7.8412
0.65	0.0838	7.1389	7.1392	7.1454	7.1570	7.1690	7.1819	7.1962
0.70	0.0598	0.2874	6.5275	6.5337	6.5471	6.5559	6.5674	6.5815
0.75	0.0370	0.1823	0.3717	6.0492	6.0546	6.0648	6.0732	6.0851
0.80	0.0279	0.1385	0.2789	0.5904	5.4371	5.4431	5.4483	5.4559
0.85	0.0213	0.1066	0.2143	0.4435	4.8375	4.8332	4.8366	4.8449
0.90	0.0166	0.0830	0.1669	0.3401	0.7575	4.2359	4.2279	4.2277
0.93	0.0143	0.0717	0.1440	0.2927	0.6261	1.1494	3.8539	3.8411
0.95	0.0130	0.0652	0.1310	0.2651	0.5579	0.9514	3.5967	3.5713
0.97	0.0118	0.0593	0.1190	0.2407	0.4999	0.8149	1.4327	3.2740
0.98	0.0113	0.0566	0.1134	0.2295	0.4742	0.7609	1.2097	3.1017
0.99	0.0108	0.0540	0.1085	0.2188	0.4504	0.7122	1.0693	2.8952
1.00	0.0103	0.0516	0.1034	0.2089	0.4275	0.6688	0.9683	1.2925
1.01	0.0099	0.0493	0.0987	0.1988	0.4064	0.6304	0.8895	1.2338
1.02	0.0094	0.0471	0.0945	0.1900	0.3866	0.5949	0.8240	1.0703
1.05	0.0082	0.0410	0.0821	0.1652	0.3329	0.5040	0.6742	0.8216
1.1	0.0066	0.0328	0.0657	0.1310	0.2617	0.3890	0.5072	0.6013
1.15	0.0053	0.0262	0.0524	0.1043	0.2062	0.3031	0.3901	0.4600
1.2	0.0042	0.0210	0.0418	0.0828	0.1626	0.2367	0.3015	0.3524
1.3	0.0027	0.0133	0.0263	0.0519	0.0997	0.1423	0.1782	0.2053
1.4	0.0021	0.0102	0.0201	0.0393	0.0738	0.1032	0.1268	0.1438
1.5	0.0013	0.0064	0.0124	0.0240	0.0439	0.0592	0.0703	0.0768
1.6	0.0007	0.0036	0.0069	0.0129	0.0222	0.0278	0.0306	0.0293
1.7	0.0003	0.0016	0.0029	0.0049	0.0065	0.0054	0.0013	-0.0051
1.8	0.0000	0.0000	-0.0002	-0.0011	-0.0049	-0.0114	-0.0201	-0.0304
1.9	-0.0002	-0.0011	-0.0024	-0.0056	-0.0135	-0.0235	-0.0359	-0.0499
2.0	-0.0004	-0.0020	-0.0042	-0.0090	-0.0204	-0.0331	-0.0480	-0.0640
2.2	-0.0006	-0.0032	-0.0066	-0.0137	-0.0292	-0.0459	-0.0640	-0.0833
2.4	-0.0008	-0.0039	-0.0081	-0.0165	-0.0345	-0.0537	-0.0736	-0.0946
2.6	-0.0009	-0.0044	-0.0089	-0.0182	-0.0377	-0.0580	-0.0792	-0.101
2.8	-0.0009	-0.0046	-0.0094	-0.0191	-0.0394	-0.0605	-0.0822	-0.1044
3.0	-0.0009	-0.0048	-0.0096	-0.0196	-0.0403	-0.0616	-0.0835	-0.1059
3.5	-0.0010	-0.0049	-0.0099	-0.0201	-0.0411	-0.0625	-0.0844	-0.1065
4.0	-0.0009	-0.0048	-0.0097	-0.0197	-0.0401	-0.0610	-0.0822	-0.1037

Table 5.18 Values of $(H^0 - H)/RT_c\big|^{(1)}$ **(continued)**

T_r / p_r	1.2	1.5	2	3	5	7	10
0.55	8.5245	8.5468	8.5808	8.6549	8.8003	8.9442	9.1605
0.60	7.8517	7.8709	7.9093	7.9809	8.1230	8.2663	8.4858
0.65	7.2070	7.2284	7.2626	7.3343	7.4760	7.6212	7.8406
0.70	6.5929	6.6097	6.6418	6.7100	6.8540	7.0076	7.2268
0.75	6.0911	6.1103	6.1400	6.2048	6.3441	6.4896	6.7011
0.80	5.4655	5.4813	5.5127	5.5824	5.7301	5.8804	6.1016
0.85	4.8544	4.8687	4.9010	4.9792	5.1444	5.3060	5.5395
0.90	4.2325	4.2546	4.2914	4.3893	4.5771	4.7604	5.0096
0.93	3.8521	3.8746	3.9213	4.0346	4.2459	4.4419	4.7046
0.95	3.5809	3.6074	3.6692	3.7971	4.0322	4.2381	4.5117
0.97	3.2805	3.3264	3.4053	3.5588	3.8175	4.0357	4.3202
0.98	3.1152	3.1754	3.2713	3.4395	3.7138	3.9358	4.2218
0.99	2.9321	3.0142	3.1320	3.3223	3.6090	3.8400	4.1289
1.00	2.7093	2.8429	2.9901	3.2021	3.5001	3.7400	4.0380
1.01	2.3975	2.6531	2.8415	3.0776	3.4016	3.6431	3.9483
1.02	1.8007	2.4361	2.6887	2.9559	3.2971	3.5487	3.8573
1.05	0.7500	1.4852	2.1769	2.5831	2.9890	3.2701	3.5944
1.10	0.6358	0.4550	1.1626	1.9446	2.4945	2.8205	3.1777
1.15	0.4951	0.4472	0.4381	1.3189	2.0313	2.4022	2.7905
1.2	0.3819	0.3741	0.2741	0.7848	1.5935	2.0108	2.4267
1.3	0.2215	0.2220	0.1690	0.2059	0.8712	1.3331	1.7870
1.4	0.1537	0.1538	0.1215	0.05740	0.4006	0.8151	1.2653
1.5	0.0782	0.0717	0.0412	-0.0404	0.0942	0.4167	0.8379
1.6	0.0248	0.0116	-0.0231	-0.1071	-0.0973	0.1261	0.5004
1.7	-0.0145	-0.0327	-0.0720	-0.1610	-0.2203	-0.0858	0.2219
1.8	-0.0435	-0.0654	-0.1092	-0.2041	-0.3036	-0.2385	-0.0023
1.9	-0.0651	-0.0906	-0.1379	-0.2364	-0.3646	-0.3540	-0.1806
2.0	-0.0815	-0.1093	-0.1597	-0.2624	-0.4128	-0.4374	-0.3242
2.2	-0.1034	-0.1345	-0.1886	-0.2976	-0.4730	-0.5528	-0.5359
2.4	-0.1160	-0.1493	-0.2054	-0.3181	-0.5090	-0.6269	-0.6725
2.6	-0.1230	-0.1571	-0.2148	-0.3293	-0.5304	-0.6709	-0.7673
2.8	-0.1268	-0.1613	-0.2194	-0.3351	-0.5420	-0.6980	-0.8315
3.0	-0.1284	-0.1630	-0.2211	-0.3368	-0.5474	-0.7143	-0.8760
3.5	-0.1289	-0.1627	-0.2195	-0.3331	-0.5451	-0.7257	-0.9286
4.0	-0.1254	-0.1582	-0.2134	-0.3237	-0.5348	-0.7231	-0.9518

Table 5.19 Values of $(H^0 - H)/RT_c\Big|^{(2)}$

T_r / p_r	0.01	0.05	0.1	0.2	0.4	0.6	0.8	1
0.55	104.85	-108.91	-109.46	-109.44	-110.01	-110.21	-110.32	-110.76
0.60	6.2699	-82.164	-82.322	-82.369	-82.619	-82.603	-82.919	-83.351
0.65	2.9662	-57.259	-56.994	-57.165	-57.121	-57.282	-57.598	-57.753
0.70	1.5255	16.477	-33.644	-33.814	-34.270	-34.221	-34.389	-34.662
0.75	1.4664	10.393	31.506	-21.986	-21.786	-22.051	-21.987	-22.256
0.80	0.9103	5.9518	15.785	59.002	2.9423	2.5615	2.5893	2.5233
0.85	0.5749	3.5595	8.9088	26.557	26.132	25.994	25.722	25.365
0.90	0.3625	2.1589	5.2222	14.445	52.939	48.516	47.772	47.283
0.93	0.2750	1.6018	3.8070	10.203	33.809	98.820	61.239	60.536
0.95	0.2239	1.3068	3.0713	8.1273	25.813	61.972	70.983	69.212
0.97	0.1832	1.0573	2.4783	6.4457	19.901	44.171	109.09	79.012
0.98	0.1644	0.9501	2.2286	5.7430	17.480	37.789	77.758	84.452
0.99	0.1474	0.8515	1.9684	5.1111	15.353	32.755	62.551	90.753
1.00	0.1330	0.7566	1.7610	4.5112	13.509	28.355	51.808	101.656
1.01	0.1162	0.6739	1.5722	4.0127	11.868	24.545	43.984	71.676
1.02	0.1052	0.5956	1.3800	3.5462	10.418	21.323	37.319	59.384
1.05	0.0689	0.3964	0.9258	2.3620	6.9540	13.955	23.717	36.436
1.10	0.0224	0.1442	0.3503	0.9992	3.1155	6.4077	10.878	16.685
1.15	-0.0097	-0.0298	-0.0287	0.0973	0.7646	2.0240	3.8982	6.3772
1.2	-0.0338	-0.1577	-0.2964	-0.5120	-0.7241	-0.6144	-0.1460	0.7282
1.3	-0.0658	-0.3250	-0.6388	-1.2471	-2.3729	-3.3748	-4.2145	-4.8201
1.4	-0.2030	-1.0088	-1.9960	-3.9191	-7.5515	-10.856	-13.811	-16.412
1.5	-0.2015	-0.9987	-1.9805	-3.8930	-7.5357	-10.827	-13.862	-16.590
1.6	-0.1960	-0.9709	-1.9232	-3.7812	-7.2824	-10.518	-13.520	-16.190
1.7	-0.1884	-0.9347	-1.8495	-3.6266	-6.9835	-10.079	-12.911	-15.511
1.8	-0.1794	-0.8900	-1.7605	-3.4559	-6.6476	-9.5714	-12.255	-14.744
1.9	-0.1703	-0.8415	-1.6705	-3.2717	-6.2982	-9.0613	-11.590	-13.890
2.0	-0.1608	-0.7968	-1.5760	-3.0842	-5.9224	-8.5329	-10.890	-13.064
2.2	-0.1423	-0.7031	-1.3922	-2.7230	-5.2122	-7.4877	-9.5816	-11.426
2.4	-0.1248	-0.6172	-1.2164	-2.3832	-4.5498	-6.5216	-8.3162	-9.9113
2.6	-0.1088	-0.5375	-1.0611	-2.0706	-3.9410	-5.6353	-7.1638	-8.5325
2.8	-0.0940	-0.4645	-0.9159	-1.7842	-3.3882	-4.8290	-6.1149	-7.2671
3.0	-0.0807	-0.3979	-0.7836	-1.5255	-2.8794	-4.0926	-5.1655	-6.1098
3.5	-0.0530	-0.2599	-0.5095	-0.9820	-1.8253	-2.5485	-3.1569	-3.6603
4.0	-0.0302	-0.1461	-0.2835	-0.5345	-0.9540	-1.2705	-1.4938	-1.6293

Table 5.19 Values of $(H^0 - H)/RT_c\big|^{(2)}$ **(continued)**

T_r / p_r	1.2	1.5	2	3	5	7	10
0.55	-110.83	-111.32	-111.96	-112.86	-114.81	-115.85	-116.35
0.60	-83.200	-83.621	-84.231	-85.078	-86.152	-86.611	-87.031
0.65	-57.833	-58.034	-58.603	-59.450	-60.008	-60.564	-60.288
0.70	-34.684	-34.637	-34.837	-35.407	-36.353	-37.040	-36.573
0.75	-22.029	-22.452	-22.489	-22.723	-22.785	-22.393	-20.196
0.80	2.4525	2.2462	1.7367	1.1422	0.6222	0.9568	2.9905
0.85	25.005	24.713	23.997	22.504	21.176	21.402	23.123
0.90	46.756	45.547	44.381	41.682	39.287	38.549	40.317
0.93	59.165	57.680	55.599	52.104	48.984	48.170	49.495
0.95	67.614	65.616	62.606	58.877	54.605	53.662	54.905
0.97	76.546	73.355	69.769	64.714	60.058	58.773	59.995
0.98	81.180	77.154	72.961	67.671	62.592	61.260	62.418
0.99	85.515	81.069	76.159	70.129	64.892	63.361	64.842
1.00	90.359	84.560	78.963	72.803	67.558	65.878	66.970
1.01	94.465	87.793	81.791	75.214	69.162	67.936	68.903
1.02	89.056	90.588	84.232	77.374	71.308	69.784	71.217
1.05	52.043	81.516	88.577	82.538	76.742	74.992	76.470
1.10	23.871	40.231	71.520	84.568	82.115	81.268	83.581
1.15	9.9468	17.823	39.330	74.949	82.755	84.431	88.353
1.2	2.2797	5.8263	18.747	55.074	78.844	84.469	90.699
1.3	-5.0984	-4.7967	-0.8904	18.401	58.629	74.942	88.689
1.4	-18.608	-20.964	-22.123	-13.907	25.004	52.421	76.808
1.5	-18.956	-21.864	-24.757	-22.588	3.3433	32.053	61.793
1.6	-18.589	-21.581	-25.048	-26.098	-9.5882	15.269	46.115
1.7	-17.833	-20.791	-24.432	-26.945	-16.561	3.8279	33.527
1.8	-16.929	-19.813	-23.416	-26.586	-20.149	-3.9512	23.709
1.9	-15.953	-18.695	-22.169	-25.661	-21.559	-8.0561	16.651
2.0	-15.012	-17.565	-20.845	-24.369	-21.499	-10.708	11.799
2.2	-13.112	-15.351	-18.216	-21.394	-19.814	-11.512	7.4595
2.4	-11.389	-13.232	-15.702	-18.331	-16.915	-9.6341	6.8378
2.6	-9.7774	-11.331	-13.350	-15.379	-13.580	-6.7141	8.6041
2.8	-8.2869	-9.5805	-11.164	-12.545	-10.129	-3.2523	11.589
3.0	-6.9358	-7.9558	-9.1311	-9.8924	-6.7122	0.4376	15.262
3.5	-4.0639	-4.4939	-4.7940	-4.0201	1.3310	9.8982	25.948
4.0	-1.6786	-1.6066	-1.1302	0.9571	8.4313	18.655	36.576

Table 5.20 Values of $(S^0 - S)/R\big|^{(0)}$

T_r / p_r	0.01	0.05	0.1	0.2	0.4	0.6	0.8	1
0.55	0.0299	8.2034	7.5134	6.8268	6.1477	5.7555	5.4810	5.2711
0.60	0.0223	7.9600	7.2699	6.5855	5.9075	5.5170	5.2441	5.0355
0.65	0.0190	7.7285	7.0400	6.3559	5.6800	5.2915	5.0204	4.8141
0.70	0.0158	0.0844	6.8139	6.1313	5.4584	5.0725	4.8046	4.6006
0.75	0.0135	0.0707	0.1508	5.8993	5.2310	4.8490	4.5847	4.3837
0.80	0.0113	0.0588	0.1238	0.2784	5.0024	4.6269	4.3682	4.1721
0.85	0.0097	0.0497	0.1034	0.2259	4.7492	4.3852	4.1362	3.9489
0.90	0.0083	0.0425	0.0877	0.1879	0.4487	4.0975	3.8702	3.6996
0.93	0.0076	0.0390	0.0801	0.1698	0.3932	0.7428	3.6723	3.5234
0.95	0.0072	0.0368	0.0755	0.1593	0.3628	0.6575	3.5004	3.3851
0.97	0.0068	0.0348	0.0713	0.1497	0.3364	0.5919	1.0531	3.2110
0.98	0.0067	0.0339	0.0693	0.1453	0.3245	0.5642	0.9594	3.0978
0.99	0.0065	0.0330	0.0675	0.1411	0.3133	0.5390	0.8879	2.9425
1.00	0.0063	0.0322	0.0657	0.1371	0.3028	0.5161	0.8297	2.3027
1.01	0.0062	0.0314	0.0639	0.1332	0.2929	0.4950	0.7806	1.3614
1.02	0.0060	0.0306	0.0623	0.1296	0.2835	0.4755	0.7382	1.1961
1.05	0.0056	0.0284	0.0577	0.1194	0.2582	0.4251	0.6376	0.9375
1.10	0.0050	0.0252	0.0510	0.1050	0.2236	0.3601	0.5217	0.7204
1.15	0.0045	0.0225	0.0455	0.0932	0.1960	0.3109	0.4409	0.5908
1.2	0.0040	0.0202	0.0408	0.0833	0.1736	0.2721	0.3805	0.5005
1.3	0.0033	0.0166	0.0334	0.0677	0.1393	0.2151	0.2953	0.3804
1.4	0.0027	0.0137	0.0276	0.0557	0.1135	0.1737	0.2361	0.3008
1.5	0.0023	0.0116	0.0233	0.0469	0.0951	0.1444	0.1950	0.2466
1.6	0.0020	0.0099	0.0199	0.0400	0.0808	0.1222	0.1642	0.2067
1.7	0.0017	0.0086	0.0172	0.0346	0.0695	0.1048	0.1403	0.1761
1.8	0.0015	0.0075	0.0150	0.0301	0.0604	0.0909	0.1214	0.1519
1.9	0.0013	0.0066	0.0132	0.0265	0.0530	0.0796	0.1061	0.1325
2.0	0.0012	0.0059	0.0117	0.0234	0.0469	0.0702	0.0935	0.1167
2.2	0.0009	0.0047	0.0094	0.0187	0.0373	0.0558	0.0742	0.0924
2.4	0.0008	0.0038	0.0076	0.0152	0.0304	0.0454	0.0602	0.0750
2.6	0.0006	0.0032	0.0063	0.0126	0.0252	0.0376	0.0498	0.0619
2.8	0.0005	0.0027	0.0053	0.0106	0.0211	0.0315	0.0418	0.0520
3.0	0.0005	0.0023	0.0045	0.0090	0.0180	0.0268	0.0355	0.0442
3.5	0.0003	0.0016	0.0031	0.0062	0.0123	0.0183	0.0243	0.0302
4.0	0.0002	0.0011	0.0023	0.0045	0.0089	0.0133	0.0177	0.0220

Table 5.20 Values of $(S^0 - S)/R\big|^{(0)}$ **(continued)**

T_r / p_r	1.2	1.5	2	3	5	7	10
0.55	5.1013	4.8978	4.6412	4.2952	3.8963	3.6613	3.4434
0.60	4.8673	4.6654	4.4123	4.0715	3.6801	3.4521	3.2406
0.65	4.6474	4.4480	4.1984	3.8646	3.4834	3.2619	3.0584
0.70	4.4364	4.2401	3.9950	3.6695	3.3004	3.0873	2.8923
0.75	4.2226	4.0307	3.7927	3.4775	3.1219	2.9185	2.7332
0.80	4.0156	3.8301	3.5999	3.2982	2.9601	2.7674	2.5927
0.85	3.8003	3.6232	3.4061	3.1215	2.8043	2.6242	2.4607
0.90	3.5639	3.4032	3.2049	2.9450	2.6537	2.4875	2.3364
0.93	3.4036	3.2590	3.0776	2.8374	2.5645	2.4081	2.2645
0.95	3.2825	3.1541	2.9893	2.7649	2.5063	2.3561	2.2185
0.97	3.1431	3.0406	2.8965	2.6910	2.4483	2.3051	2.1731
0.98	3.0629	2.9790	2.8483	2.6540	2.4193	2.2799	2.1506
0.99	2.9716	2.9133	2.7988	2.6164	2.3904	2.2550	2.1286
1.00	2.8642	2.8428	2.7476	2.5783	2.3616	2.2302	2.1067
1.01	2.7294	2.7658	2.6945	2.5404	2.3331	2.2051	2.0848
1.02	2.5398	2.6814	2.6384	2.5017	2.3047	2.1805	2.0635
1.05	1.4769	2.3522	2.4568	2.3825	2.2191	2.1085	2.0002
1.10	0.9778	1.5279	2.0854	2.1723	2.0777	1.9906	1.8991
1.15	0.7661	1.0890	1.6451	1.9478	1.9374	1.8763	1.8029
1.2	0.6342	0.8626	1.2811	1.7144	1.7986	1.7654	1.7102
1.3	0.4705	0.6145	0.8691	1.2851	1.5304	1.5547	1.5375
1.4	0.3677	0.4715	0.6496	0.9735	1.2847	1.3614	1.3804
1.5	0.2993	0.3796	0.5147	0.7668	1.0750	1.1872	1.2373
1.6	0.2496	0.3143	0.4220	0.6241	0.9054	1.0352	1.1092
1.7	0.2119	0.2656	0.3542	0.5209	0.7701	0.9050	0.9943
1.8	0.1824	0.2280	0.3027	0.4431	0.6625	0.7949	0.8938
1.9	0.1589	0.1980	0.2622	0.3825	0.5761	0.7020	0.8053
2.0	0.1397	0.1738	0.2295	0.3341	0.5058	0.6238	0.7277
2.2	0.1105	0.1372	0.1806	0.2621	0.3995	0.5010	0.6000
2.4	0.0895	0.1110	0.1459	0.2115	0.3240	0.4108	0.5015
2.6	0.0739	0.0916	0.1203	0.1743	0.2681	0.3427	0.4244
2.8	0.0620	0.0768	0.1008	0.1461	0.2255	0.2900	0.3632
3.0	0.0527	0.0652	0.0856	0.1241	0.1921	0.2484	0.3139
3.5	0.0360	0.0446	0.0585	0.0850	0.1324	0.1728	0.2221
4.0	0.0262	0.0325	0.0426	0.0620	0.0970	0.1274	0.1655

Table 5.21 Values of $(S^0 - S)/R\big|^{(1)}$

T_r / p_r	0.01	0.05	0.1	0.2	0.4	0.6	0.8
0.55	-0.3543	10.202	10.207	10.218	10.216	10.220	10.220
0.60	0.1727	9.0496	9.0665	9.0603	9.0543	9.0548	9.0553
0.65	0.1058	8.0351	8.035	8.0304	8.0309	8.0281	8.0268
0.70	0.0702	0.3366	7.1277	7.1265	7.1207	7.1190	7.1127
0.75	0.0409	0.2013	0.4114	6.4906	6.4721	6.4663	6.4593
0.80	0.0291	0.1444	0.2910	0.6206	5.6685	5.6599	5.6499
0.85	0.0211	0.1058	0.2127	0.4424	4.9442	4.9272	4.9157
0.90	0.0157	0.0787	0.1583	0.3240	0.7321	4.2443	4.2137
0.93	0.0132	0.0664	0.1336	0.2721	0.5881	1.1035	3.8034
0.95	0.0119	0.0594	0.1195	0.2431	0.5155	0.8937	3.5351
0.97	0.0106	0.0533	0.1072	0.2178	0.4554	0.7505	1.3565
0.98	0.0101	0.0505	0.1015	0.2057	0.4288	0.6944	1.1280
0.99	0.0095	0.0479	0.0964	0.1950	0.4042	0.6454	0.9856
1.00	0.0091	0.0455	0.0914	0.1846	0.3813	0.6023	0.8850
1.01	0.0086	0.0431	0.0867	0.1753	0.3605	0.5636	0.8044
1.02	0.0082	0.0410	0.0823	0.1662	0.3406	0.5287	0.7396
1.05	0.0070	0.0352	0.0706	0.1423	0.2891	0.4411	0.5955
1.10	0.0055	0.0275	0.0551	0.1110	0.2226	0.3337	0.4397
1.15	0.0043	0.0217	0.0434	0.0869	0.1733	0.2572	0.3349
1.2	0.0034	0.0172	0.0344	0.0685	0.1361	0.2009	0.2599
1.3	0.0022	0.0110	0.0220	0.0436	0.0858	0.1249	0.1608
1.4	0.0017	0.0084	0.0166	0.0328	0.0637	0.0922	0.1174
1.5	0.0011	0.0057	0.0114	0.0223	0.0429	0.0618	0.0785
1.6	0.0008	0.0039	0.0077	0.0152	0.0292	0.0416	0.0523
1.7	0.0005	0.0027	0.0053	0.0104	0.0197	0.0281	0.0352
1.8	0.0004	0.0018	0.0036	0.0069	0.0130	0.0183	0.0227
1.9	0.0002	0.0012	0.0023	0.0045	0.0083	0.0115	0.0140
2.0	0.0002	0.0007	0.0014	0.0028	0.0049	0.0067	0.0077
2.2	0.0000	0.0002	0.0003	0.0005	0.0006	0.0005	0.0000
2.4	0.0000	-0.0002	-0.0003	-0.0007	-0.0016	-0.0028	-0.0041
2.6	-0.0001	-0.0003	-0.0007	-0.0014	-0.0029	-0.0046	-0.0063
2.8	-0.0001	-0.0004	-0.0009	-0.0018	-0.0035	-0.0055	-0.0074
3.0	-0.0001	-0.0005	-0.0009	-0.0019	-0.0039	-0.0059	-0.0079
3.5	-0.0001	-0.0004	-0.0009	-0.0017	-0.0034	-0.0052	-0.0069
4.0	-0.0001	-0.0004	-0.0008	-0.0016	-0.0032	-0.0048	-0.0063

Table 5.21 Values of $(S^0 - S)/R\big|^{(1)}$ **(continued)**

T_r / p_r	1.2	1.5	2	3	5	7	10
0.55	10.226	10.228	10.236	10.255	10.297	10.359	10.456
0.60	9.0581	9.0582	9.0622	9.0806	9.1218	9.1785	9.2836
0.65	8.0259	8.0249	8.0298	8.0403	8.0845	8.1432	8.2513
0.70	7.1093	7.1066	7.1137	7.1211	7.1637	7.2268	7.3409
0.75	6.4541	6.4543	6.4535	6.4628	6.5133	6.5786	6.6912
0.80	5.6470	5.6417	5.6434	5.6599	5.7204	5.7930	5.9170
0.85	4.8993	4.8996	4.9013	4.9284	5.0067	5.0953	5.2334
0.90	4.1938	4.1926	4.2069	4.2494	4.3606	4.4694	4.6267
0.93	3.7720	3.7755	3.8030	3.8650	4.0010	4.1259	4.2959
0.95	3.4865	3.4949	3.5307	3.6162	3.7696	3.9065	4.0853
0.97	3.1777	3.2019	3.2593	3.3687	3.5488	3.6979	3.8874
0.98	3.0078	3.0474	3.1207	3.2443	3.4417	3.5958	3.7902
0.99	2.8189	2.8841	2.9796	3.1230	3.3340	3.4943	3.6973
1.00	2.5962	2.7102	2.8332	3.0042	3.2297	3.3971	3.6020
1.01	2.2874	2.5226	2.6833	2.8799	3.1270	3.3028	3.5145
1.02	1.7010	2.3090	2.5375	2.7602	3.0221	3.2075	3.4244
1.05	0.6673	1.3909	2.0442	2.3971	2.7282	2.9364	3.1697
1.10	0.5634	0.4273	1.0995	1.8065	2.2698	2.5202	2.7823
1.15	0.4384	0.4163	0.4518	1.2473	1.8538	2.1464	2.4326
1.2	0.3430	0.3538	0.3156	0.7939	1.4857	1.8153	2.1282
1.3	0.2139	0.2330	0.2289	0.3301	0.8985	1.2729	1.6143
1.4	0.1560	0.1734	0.1821	0.2040	0.5367	0.8728	1.2175
1.5	0.1038	0.1164	0.1262	0.1370	0.3258	0.5986	0.9227
1.6	0.0693	0.0778	0.0851	0.0924	0.2024	0.4093	0.6978
1.7	0.0456	0.0507	0.0555	0.0597	0.1273	0.2815	0.5342
1.8	0.0290	0.0320	0.0340	0.0354	0.0789	0.1924	0.4042
1.9	0.0171	0.0187	0.0184	0.0177	0.0456	0.1306	0.3078
2.0	0.0089	0.0089	0.0073	0.0046	0.0223	0.0863	0.2338
2.2	-0.0015	-0.0032	-0.0067	-0.0123	-0.0069	0.0311	0.1338
2.4	-0.0071	-0.0099	-0.0141	-0.0215	-0.0228	0.0000	0.0731
2.6	-0.0100	-0.0129	-0.0177	-0.0257	-0.0315	-0.0177	0.0352
2.8	-0.0114	-0.0145	-0.0195	-0.0280	-0.0359	-0.0276	0.0115
3.0	-0.0120	-0.0150	-0.0199	-0.0285	-0.0376	-0.0336	-0.0036
3.5	-0.0103	-0.0127	-0.0166	-0.0236	-0.0316	-0.0309	-0.0138
4.0	-0.0093	-0.0115	-0.0150	-0.0211	-0.0290	-0.0303	-0.0194

Table 5.22 Values of $(S^0 - S)/R|^{(2)}$

T_r / p_r	0.01	0.05	0.1	0.2	0.4	0.6	0.8	1
0.55	186.24	-169.73	-170.25	-172.81	-174.25	-176.14	-177.50	-178.99
0.60	9.5323	-123.87	-126.50	-126.47	-127.23	-128.71	-130.34	-131.79
0.65	4.2128	-84.164	-84.620	-85.081	-86.714	-88.003	-89.441	-90.505
0.70	2.0750	22.0914	-49.973	-50.955	-52.111	-53.573	-54.895	-56.081
0.75	1.7837	12.638	38.700	-40.068	-38.835	-39.510	-40.750	-41.844
0.80	1.0573	6.9136	18.315	69.037	-5.768	-7.1526	-8.4730	-9.9882
0.85	0.6523	3.9933	9.9691	29.450	22.509	20.934	19.031	17.702
0.90	0.4101	2.3976	5.7254	15.553	55.998	45.9740	44.654	42.598
0.93	0.3112	1.7860	4.1578	10.900	35.171	102.410	59.349	56.811
0.95	0.2590	1.4729	3.3986	8.6580	26.659	63.1810	69.404	66.298
0.97	0.2180	1.2150	2.7553	6.9233	20.469	44.479	109.36	76.584
0.98	0.1976	1.1016	2.5059	6.2295	18.031	38.089	77.273	82.385
0.99	0.1813	1.0026	2.2463	5.5744	15.881	32.837	61.576	88.413
1.00	0.1651	0.9095	2.0378	4.9973	14.026	28.422	50.932	99.583
1.01	0.1503	0.8266	1.8421	4.4673	12.366	24.683	43.070	69.409
1.02	0.1369	0.7500	1.6611	4.0034	10.958	21.524	36.837	57.565
1.05	0.1027	0.5541	1.2157	2.8807	7.5744	14.377	23.551	35.147
1.10	0.0601	0.3204	0.6928	1.5744	4.0288	7.3326	11.523	16.725
1.15	0.0312	0.1609	0.3507	0.7821	1.9297	3.4623	5.3431	7.7034
1.2	0.0109	0.0562	0.1179	0.2802	0.6671	1.1780	1.8611	2.8044
1.3	-0.0150	-0.0752	-0.1582	-0.3220	-0.6828	-1.0324	-1.4056	-1.6722
1.4	-0.0987	-0.4929	-0.9780	-1.9316	-3.7764	-5.5247	-7.1199	-8.5611
1.5	-0.0974	-0.4856	-0.9684	-1.9159	-3.7422	-5.5124	-7.1643	-8.6958
1.6	-0.0939	-0.4671	-0.9281	-1.8452	-3.6166	-5.3166	-6.8977	-8.3845
1.7	-0.0893	-0.4448	-0.8869	-1.7516	-3.4356	-5.0552	-6.5761	-8.0020
1.8	-0.0842	-0.4196	-0.8351	-1.6506	-3.2350	-4.7520	-6.1884	-7.5327
1.9	-0.0793	-0.3948	-0.7850	-1.5531	-3.0426	-4.4629	-5.8194	-7.1163
2.0	-0.0746	-0.3705	-0.7380	-1.4598	-2.8544	-4.1915	-5.4581	-6.6624
2.2	-0.0656	-0.3266	-0.6498	-1.2842	-2.5151	-3.6955	-4.8223	-5.9029
2.4	-0.0580	-0.2889	-0.5748	-1.1357	-2.2304	-3.2773	-4.2809	-5.2450
2.6	-0.0516	-0.2570	-0.5114	-1.0104	-1.9853	-2.9213	-3.8243	-4.6901
2.8	-0.0462	-0.2298	-0.4575	-0.9058	-1.7814	-2.6225	-3.4335	-4.2175
3.0	-0.0415	-0.2067	-0.4119	-0.8165	-1.6044	-2.3676	-3.1035	-3.8185
3.5	-0.0322	-0.1603	-0.3195	-0.6343	-1.2503	-1.8488	-2.4297	-2.9971
4.0	-0.0261	-0.1301	-0.2592	-0.5150	-1.0171	-1.5065	-1.9852	-2.4516

Table 5.22 Values of $(S^0 - S)/R\big|^{(2)}$ **(continued)**

T_r / p_r	1.2	1.5	2	3	5	7	10
0.55	-180.86	-183.25	-187.44	-195.38	-210.91	-226.33	-247.88
0.60	-133.32	-135.45	-139.24	-147.02	-161.29	-175.80	-196.96
0.65	-92.205	-94.304	-98.096	-105.06	-119.72	-133.57	-154.40
0.70	-57.648	-59.521	-63.767	-70.210	-84.239	-98.207	-118.74
0.75	-43.357	-45.693	-49.380	-56.273	-70.772	-84.346	-103.76
0.80	-11.950	-13.923	-17.717	-25.714	-40.593	-54.217	-73.581
0.85	15.674	13.192	9.1092	0.5019	-15.296	-29.521	-49.194
0.90	40.561	37.467	32.091	22.524	5.3156	-9.6224	-29.533
0.93	54.462	50.600	44.402	33.890	15.9540	0.3384	-19.567
0.95	63.060	59.043	52.120	40.497	22.155	6.4138	-13.536
0.97	72.286	66.908	59.266	46.897	27.458	11.839	-8.5085
0.98	77.000	70.831	62.723	50.046	29.925	14.157	-5.8138
0.99	81.829	74.887	65.850	52.762	32.525	16.762	-3.7992
1.00	86.296	78.571	69.103	55.055	34.738	18.782	-1.1552
1.01	90.410	81.743	72.045	57.613	36.832	20.791	0.5719
1.02	85.093	84.334	73.966	59.708	39.034	22.989	2.6971
1.05	49.474	75.746	78.280	64.965	43.980	27.881	8.1234
1.10	23.212	36.786	62.314	66.705	49.047	33.772	14.665
1.15	10.663	17.004	34.163	58.228	49.734	36.704	19.084
1.2	4.0664	7.0038	16.384	41.213	46.065	36.485	20.858
1.3	-1.8139	-1.6090	0.4783	11.667	30.618	29.018	19.500
1.4	-9.7969	-11.239	-12.229	-8.3838	9.4724	16.590	14.158
1.5	-10.054	-11.844	-13.964	-14.453	-5.2122	2.2367	3.5424
1.6	-9.8090	-11.692	-14.200	-16.675	-13.755	-8.3556	-6.0625
1.7	-9.3631	-11.213	-13.841	-17.223	-18.011	-15.497	-14.074
1.8	-8.8320	-10.656	-13.256	-17.034	-20.003	-19.706	-19.486
1.9	-8.3244	-10.0494	-12.556	-16.526	-20.731	-22.090	-23.362
2.0	-7.8479	-9.4569	-11.899	-15.849	-20.821	-23.276	-25.791
2.2	-6.9432	-8.3995	-10.618	-14.445	-19.993	-23.756	-27.998
2.4	-6.1721	-7.4777	-9.5317	-13.103	-18.740	-22.979	-28.228
2.6	-5.5257	-6.7227	-8.5966	-11.924	-17.370	-21.829	-27.499
2.8	-4.9770	-6.0705	-7.7820	-10.866	-16.110	-20.540	-26.397
3.0	-4.5115	-5.5110	-7.0884	-9.9629	-14.932	-19.271	-25.155
3.5	-3.5479	-4.3495	-5.6268	-7.9912	-12.236	-16.101	-21.541
4.0	-2.9108	-3.5765	-4.6478	-6.6576	-10.330	-13.754	-18.749

Table 5.23 Values of $\left[\log(f/p)\right]^{(0)}$

T_r / p_r	0.01	0.05	0.1	0.2	0.4	0.6	0.8	1
0.55	-0.0074	-0.6019	-0.8988	-1.1914	-1.4750	-1.6339	-1.7418	-1.8210
0.60	-0.0059	-0.2598	-0.5567	-0.8493	-1.1340	-1.2939	-1.4023	-1.4833
0.65	-0.0048	0.02240	-0.2745	-0.5678	-0.8531	-1.0142	-1.1229	-1.2047
0.70	-0.0039	-0.0199	-0.0401	-0.3335	-0.6195	-0.7805	-0.8906	-0.9725
0.75	-0.0033	-0.0165	-0.0337	-0.1376	-0.4239	-0.5853	-0.6956	-0.7781
0.80	-0.0027	-0.0137	-0.0279	-0.0579	-0.2584	-0.4200	-0.5306	-0.6133
0.85	-0.0023	-0.0115	-0.0234	-0.0480	-0.1186	-0.2802	-0.3908	-0.4734
0.90	-0.0019	-0.0098	-0.0198	-0.0404	-0.0849	-0.1621	-0.2724	-0.3548
0.93	-0.0018	-0.0089	-0.0180	-0.0366	-0.0763	-0.1210	-0.2107	-0.2927
0.95	-0.0017	-0.0084	-0.0169	-0.0343	-0.0713	-0.1121	-0.1734	-0.255
0.97	-0.0016	-0.0079	-0.0159	-0.0322	-0.0667	-0.1041	-0.1471	-0.2201
0.98	-0.0015	-0.0077	-0.0154	-0.0313	-0.0645	-0.1006	-0.1411	-0.2039
0.99	-0.0015	-0.0074	-0.0149	-0.0303	-0.0625	-0.0971	-0.1356	-0.1886
1.00	-0.0014	-0.0072	-0.0145	-0.0294	-0.0605	-0.0939	-0.1306	-0.1745
1.01	-0.0014	-0.0070	-0.0141	-0.0285	-0.0586	-0.0908	-0.1258	-0.1662
1.02	-0.0014	-0.0068	-0.0137	-0.0277	-0.0568	-0.0878	-0.1214	-0.1591
1.05	-0.0012	-0.0062	-0.0125	-0.0254	-0.0518	-0.0797	-0.1094	-0.1416
1.10	-0.0011	-0.0054	-0.0109	-0.0220	-0.0447	-0.0684	-0.0931	-0.1190
1.15	-0.0009	-0.0047	-0.0095	-0.0192	-0.0388	-0.0591	-0.0800	-0.1015
1.2	-0.0008	-0.0042	-0.0083	-0.0168	-0.0339	-0.0514	-0.0692	-0.0875
1.3	-0.0006	-0.0032	-0.0065	-0.0130	-0.0261	-0.0393	-0.0527	-0.0662
1.4	-0.0005	-0.0025	-0.0050	-0.0101	-0.0202	-0.0304	-0.0405	-0.0507
1.5	-0.0004	-0.0020	-0.0040	-0.0079	-0.0159	-0.0238	-0.0316	-0.0394
1.6	-0.0003	-0.0016	-0.0031	-0.0063	-0.0125	-0.0187	-0.0248	-0.0308
1.7	-0.0002	-0.0012	-0.0025	-0.0049	-0.0098	-0.0147	-0.0194	-0.0241
1.8	-0.0002	-0.0010	-0.0019	-0.0039	-0.0077	-0.0115	-0.0152	-0.0189
1.9	-0.0002	-0.0008	-0.0015	-0.0030	-0.0060	-0.0090	-0.0118	-0.0146
2.0	-0.0001	-0.0006	-0.0012	-0.0023	-0.0046	-0.0069	-0.0091	-0.0112
2.2	-0.0001	-0.0003	-0.0007	-0.0013	-0.0026	-0.0038	-0.0050	-0.0061
2.4	0.0000	-0.0001	-0.0003	-0.0006	-0.0011	-0.0016	-0.0021	-0.0026
2.6	0.0000	0.0000	0.0000	-0.0001	-0.0001	-0.0001	-0.0001	-0.0001
2.8	0.0000	0.0001	0.0001	0.0003	0.0006	0.0009	0.0013	0.0017
3.0	0.0000	0.0001	0.0003	0.0006	0.0011	0.0017	0.0023	0.0030
3.5	0.0000	0.0002	0.0004	0.0009	0.0018	0.0027	0.0036	0.0045
4.0	0.0001	0.0003	0.0005	0.0011	0.0021	0.0032	0.0043	0.0054

Table 5.23 Values of $\left[\log(f/p)\right]^{(0)}$ **(continued)**

T_r / p_r	1.2	1.5	2	3	5	7	10
0.55	-1.8835	-1.9547	-2.0368	-2.1281	-2.1825	-2.1630	-2.0730
0.60	-1.5461	-1.6189	-1.7034	-1.7993	-1.8634	-1.8541	-1.7797
0.65	-1.2681	-1.3419	-1.4286	-1.5286	-1.6007	-1.5996	-1.5374
0.70	-1.0371	-1.1118	-1.1995	-1.3028	-1.3813	-1.3868	-1.3358
0.75	-0.8428	-0.9181	-1.0070	-1.1126	-1.1971	-1.2083	-1.1657
0.80	-0.6782	-0.7541	-0.8441	-0.9515	-1.0397	-1.0557	-1.0209
0.85	-0.5385	-0.6144	-0.7049	-0.8135	-0.9047	-0.9247	-0.8963
0.90	-0.4197	-0.4955	-0.5860	-0.6951	-0.7887	-0.8117	-0.7887
0.93	-0.3573	-0.4329	-0.5230	-0.6322	-0.7267	-0.7512	-0.7310
0.95	-0.3192	-0.3944	-0.4843	-0.5933	-0.6882	-0.7136	-0.6952
0.97	-0.2839	-0.3585	-0.4480	-0.5566	-0.6518	-0.6781	-0.6612
0.98	-0.2672	-0.3416	-0.4307	-0.5391	-0.6344	-0.6610	-0.6448
0.99	-0.2513	-0.3251	-0.4139	-0.5221	-0.6175	-0.6443	-0.6290
1.00	-0.2361	-0.3094	-0.3977	-0.5056	-0.6010	-0.6282	-0.6134
1.01	-0.2217	-0.2943	-0.3821	-0.4896	-0.5849	-0.6124	-0.5984
1.02	-0.2082	-0.2798	-0.3670	-0.4740	-0.5693	-0.5971	-0.5837
1.05	-0.1777	-0.2404	-0.3249	-0.4302	-0.5250	-0.5534	-0.5418
1.10	-0.1465	-0.1912	-0.2651	-0.3657	-0.4587	-0.4878	-0.4787
1.15	-0.1238	-0.1588	-0.2184	-0.3109	-0.4009	-0.4301	-0.4229
1.2	-0.1061	-0.1347	-0.1830	-0.2647	-0.3504	-0.3792	-0.3735
1.3	-0.0797	-0.1002	-0.1339	-0.1945	-0.2679	-0.2947	-0.2906
1.4	-0.0608	-0.0759	-0.1005	-0.1453	-0.2045	-0.2277	-0.2240
1.5	-0.0472	-0.0586	-0.0771	-0.1106	-0.1568	-0.1757	-0.1712
1.6	-0.0368	-0.0456	-0.0597	-0.0849	-0.1202	-0.1346	-0.1287
1.7	-0.0288	-0.0355	-0.0463	-0.0654	-0.0918	-0.1019	-0.0943
1.8	-0.0224	-0.0276	-0.0358	-0.0502	-0.0694	-0.0758	-0.0663
1.9	-0.0173	-0.0213	-0.0275	-0.0381	-0.0516	-0.0547	-0.0434
2.0	-0.0132	-0.0162	-0.0208	-0.0284	-0.0373	-0.0376	-0.0246
2.2	-0.0071	-0.0086	-0.0109	-0.0142	-0.0162	-0.0121	0.0037
2.4	-0.0029	-0.0034	-0.0041	-0.0045	-0.0018	0.0053	0.0233
2.6	0.0000	0.0002	0.0007	0.0023	0.0082	0.0174	0.0369
2.8	0.0021	0.0028	0.0040	0.0071	0.0152	0.0260	0.0464
3.0	0.0036	0.0046	0.0065	0.0105	0.0202	0.0320	0.0531
3.5	0.0055	0.0069	0.0094	0.0147	0.0263	0.0392	0.0607
4.0	0.0065	0.0081	0.0110	0.0168	0.0292	0.0424	0.0636

Table 5.24 Values of $\left[\log(f/p)\right]^{(1)}$

T_r/p_r	0.01	0.05	0.1	0.2	0.4	0.6	0.8	1
0.55	-0.0233	-2.2337	-2.2328	-2.2370	-2.2484	-2.2584	-2.2670	-2.2828
0.60	-0.0157	-1.6942	-1.6934	-1.7027	-1.7124	-1.7210	-1.7325	-1.7403
0.65	-0.0100	-1.2793	-1.2825	-1.2861	-1.2976	-1.3035	-1.3149	-1.3201
0.70	-0.0066	-0.0323	-0.9556	-0.9599	-0.9687	-0.9783	-0.9837	-0.9926
0.75	-0.0037	-0.0182	-0.0367	-0.6822	-0.6946	-0.7031	-0.7111	-0.7186
0.80	-0.0025	-0.0124	-0.0249	-0.0509	-0.4885	-0.4956	-0.5026	-0.5096
0.85	-0.0017	-0.0085	-0.0171	-0.0344	-0.3241	-0.3314	-0.3383	-0.3452
0.90	-0.0012	-0.0059	-0.0117	-0.0234	-0.0476	-0.2018	-0.2092	-0.2160
0.93	-0.0009	-0.0047	-0.0093	-0.0184	-0.0370	-0.0576	-0.1466	-0.1531
0.95	-0.0008	-0.0040	-0.0079	-0.0157	-0.0312	-0.0470	-0.1096	-0.1165
0.97	-0.0007	-0.0034	-0.0067	-0.0133	-0.0262	-0.0392	-0.0528	-0.0843
0.98	-0.0006	-0.0031	-0.0063	-0.0123	-0.0240	-0.0354	-0.0467	-0.0699
0.99	-0.0006	-0.0029	-0.0057	-0.0113	-0.0219	-0.0321	-0.0416	-0.0565
1.00	-0.0005	-0.0027	-0.0053	-0.0103	-0.0200	-0.0290	-0.0373	-0.0443
1.01	-0.0005	-0.0024	-0.0048	-0.0094	-0.0182	-0.0262	-0.0333	-0.0382
1.02	-0.0005	-0.0022	-0.0044	-0.0087	-0.0166	-0.0236	-0.0293	-0.0334
1.05	-0.0003	-0.0017	-0.0034	-0.0065	-0.0122	-0.0169	-0.0205	-0.0218
1.10	-0.0002	-0.0010	-0.002	-0.0037	-0.0066	-0.0086	-0.0095	-0.0086
1.15	-0.0001	-0.0005	-0.0009	-0.0017	-0.0026	-0.0027	-0.0018	0.0003
1.2	0.0000	-0.0001	-0.0002	-0.0002	0.0003	0.0015	0.0036	0.0066
1.3	0.0001	0.0003	0.0007	0.0016	0.0039	0.0068	0.0102	0.0142
1.4	0.0001	0.0005	0.0010	0.0021	0.0048	0.0080	0.0116	0.0157
1.5	0.0001	0.0006	0.0013	0.0028	0.0060	0.0097	0.0136	0.0179
1.6	0.0001	0.0007	0.0015	0.0031	0.0066	0.0105	0.0145	0.0188
1.7	0.0002	0.0008	0.0016	0.0033	0.0068	0.0107	0.0148	0.0190
1.8	0.0002	0.0008	0.0016	0.0033	0.0069	0.0107	0.0146	0.0188
1.9	0.0002	0.0008	0.0016	0.0032	0.0067	0.0104	0.0143	0.0183
2.0	0.0001	0.0008	0.0015	0.0032	0.0065	0.0101	0.0138	0.0176
2.2	0.0001	0.0007	0.0014	0.0029	0.0060	0.0093	0.0127	0.0161
2.4	0.0001	0.0006	0.0013	0.0027	0.0055	0.0085	0.0115	0.0147
2.6	0.0001	0.0006	0.0012	0.0024	0.0050	0.0077	0.0105	0.0133
2.8	0.0001	0.0005	0.0011	0.0022	0.0046	0.0070	0.0095	0.0121
3.0	0.0001	0.0005	0.0010	0.0020	0.0041	0.0064	0.0087	0.0110
3.5	0.0001	0.0004	0.0009	0.0018	0.0036	0.0055	0.0075	0.0095
4.0	0.0001	0.0003	0.0007	0.0014	0.0030	0.0046	0.0062	0.0079

Table 5.24 Values of $\left[\log(f/p)\right]^{(1)}$ **(continued)**

T_r/p_r	1.2	1.5	2	3	5	7	10
0.55	-2.2878	-2.3043	-2.3317	-2.3824	-2.4767	-2.5683	-2.6944
0.60	-1.7525	-1.7647	-1.7890	-1.8335	-1.9177	-2.0005	-2.1116
0.65	-1.3322	-1.3438	-1.3673	-1.4060	-1.4831	-1.5556	-1.6573
0.70	-0.9979	-1.0114	-1.0327	-1.0709	-1.1413	-1.2069	-1.2966
0.75	-0.7252	-0.7366	-0.7562	-0.7868	-0.8447	-0.9008	-0.9770
0.80	-0.5163	-0.5263	-0.5427	-0.5716	-0.6254	-0.6755	-0.7435
0.85	-0.3508	-0.3608	-0.3754	-0.4035	-0.4545	-0.4972	-0.5577
0.90	-0.2221	-0.2309	-0.2447	-0.2701	-0.3150	-0.3542	-0.4076
0.93	-0.1591	-0.1677	-0.1810	-0.2046	-0.2460	-0.2829	-0.3321
0.95	-0.1229	-0.1312	-0.1436	-0.1660	-0.2052	-0.2403	-0.2868
0.97	-0.0902	-0.0983	-0.1102	-0.1314	-0.1686	-0.2011	-0.2451
0.98	-0.0757	-0.0835	-0.0950	-0.1156	-0.1513	-0.1829	-0.2256
0.99	-0.0620	-0.0695	-0.0808	-0.1003	-0.1349	-0.1655	-0.2068
1.00	-0.0496	-0.0568	-0.0674	-0.0860	-0.1193	-0.1490	-0.1890
1.01	-0.0389	-0.0449	-0.0548	-0.0726	-0.1043	-0.1331	-0.1718
1.02	-0.0298	-0.0344	-0.0431	-0.0600	-0.0906	-0.1179	-0.1555
1.05	-0.0196	-0.0099	-0.0131	-0.0263	-0.0521	-0.0765	-0.1098
1.10	-0.0060	0.0041	0.0183	0.0166	-0.0006	-0.0193	-0.0465
1.15	0.0036	0.0124	0.0312	0.0445	0.0388	0.0257	0.0047
1.2	0.0105	0.0189	0.0362	0.0609	0.0671	0.0607	0.0460
1.3	0.0188	0.0270	0.0427	0.0736	0.1011	0.1072	0.1046
1.4	0.0201	0.0275	0.0413	0.0707	0.1094	0.1261	0.1351
1.5	0.0225	0.0299	0.0431	0.0706	0.1143	0.1387	0.1569
1.6	0.0234	0.0306	0.0432	0.0693	0.1141	0.1435	0.1690
1.7	0.0235	0.0304	0.0424	0.0672	0.1115	0.1437	0.1747
1.8	0.0230	0.0297	0.0411	0.0646	0.1078	0.1414	0.1762
1.9	0.0224	0.0287	0.0396	0.0618	0.1035	0.1376	0.1750
2.0	0.0215	0.0276	0.0379	0.0589	0.0990	0.1331	0.1721
2.2	0.0197	0.0251	0.0344	0.0534	0.0903	0.1232	0.1636
2.4	0.0179	0.0228	0.0312	0.0483	0.0822	0.1134	0.1536
2.6	0.0162	0.0206	0.0282	0.0438	0.0749	0.1044	0.1435
2.8	0.0147	0.0187	0.0256	0.0398	0.0685	0.0962	0.1340
3.0	0.0134	0.0171	0.0234	0.0363	0.0629	0.0889	0.1251
3.5	0.0115	0.0147	0.0200	0.0311	0.0539	0.0766	0.1094
4.0	0.0096	0.0122	0.0167	0.0260	0.0455	0.0654	0.0948

Table 5.25 Values of $\left[\log(f/p)\right]^{(2)}$

T_r / p_r	0.01	0.05	0.1	0.2	0.4	0.6	0.8	1
0.55	-1.9316	12.126	11.767	11.402	11.020	10.442	9.8744	9.7392
0.60	-0.3931	5.4411	4.6055	4.9181	4.5611	3.9937	3.5592	3.1163
0.65	-0.1497	1.5759	1.4958	1.1207	0.8597	0.2893	-0.2733	-1.0333
0.70	-0.0461	-0.6232	-0.7084	-0.9765	-1.4644	-1.9342	-2.7382	-3.1571
0.75	-0.0764	-0.5171	-1.4271	-4.8173	-4.2150	-4.4743	-4.9193	-5.3623
0.80	-0.0335	-0.2295	-0.6174	-2.0496	-4.1024	-4.6514	-5.2184	-5.7786
0.85	-0.0102	-0.0823	-0.2357	-0.7917	-3.5958	-4.1609	-4.7390	-5.3212
0.90	0.0027	-0.0010	-0.0380	-0.2228	-1.1860	-3.3170	-3.7709	-4.3333
0.93	0.0078	0.0272	0.0311	-0.0368	-0.5118	-1.6877	-2.8462	-3.4736
0.95	0.0101	0.0420	0.0670	0.0560	-0.2195	-0.9262	-2.2697	-2.8426
0.97	0.0121	0.0535	0.0898	0.1224	-0.0046	-0.4095	-1.2888	-2.1361
0.98	0.0128	0.0583	0.1045	0.1487	0.0792	-0.2449	-0.8818	-1.7467
0.99	0.0135	0.0622	0.1123	0.1762	0.1546	-0.0677	-0.5914	-1.3531
1.00	0.0143	0.0656	0.1221	0.1952	0.2178	0.0534	-0.3241	-0.9582
1.01	0.0146	0.0686	0.1273	0.2126	0.2729	0.1656	-0.1255	-0.6236
1.02	0.0152	0.0714	0.1323	0.2321	0.3216	0.2599	0.0318	-0.3463
1.05	0.0162	0.0775	0.1473	0.2658	0.4239	0.4716	0.4108	0.2051
1.10	0.0170	0.0822	0.1596	0.2985	0.5171	0.6584	0.7275	0.6871
1.15	0.0172	0.0834	0.1613	0.3063	0.5491	0.7288	0.8471	0.9046
1.2	0.0168	0.0818	0.1590	0.3033	0.5480	0.7395	0.8739	0.9555
1.3	0.0154	0.0748	0.1459	0.2774	0.5030	0.6797	0.8085	0.8897
1.4	0.0202	0.0992	0.1946	0.3755	0.6984	0.9757	1.2016	1.3785
1.5	0.0160	0.0784	0.1534	0.2945	0.5458	0.7464	0.9102	1.0381
1.6	0.0124	0.0607	0.1180	0.2251	0.4083	0.5544	0.6662	0.7412
1.7	0.0093	0.0454	0.0879	0.1658	0.2951	0.3900	0.4532	0.4864
1.8	0.0067	0.0323	0.0624	0.1154	0.1979	0.2505	0.2750	0.2712
1.9	0.0045	0.0214	0.0405	0.0729	0.1155	0.1317	0.1232	0.0913
2.0	0.0026	0.0120	0.0219	0.0365	0.0459	0.0314	-0.0057	-0.0635
2.2	-0.0004	-0.0029	-0.0074	-0.0211	-0.0640	-0.1268	-0.2079	-0.3059
2.4	-0.0026	-0.0137	-0.0289	-0.0630	-0.1446	-0.2422	-0.3550	-0.4820
2.6	-0.0042	-0.0218	-0.0448	-0.0940	-0.2036	-0.3270	-0.4628	-0.6105
2.8	-0.0054	-0.0277	-0.0566	-0.1171	-0.2475	-0.3896	-0.5422	-0.7049
3.0	-0.0063	-0.0322	-0.0654	-0.1342	-0.2800	-0.4355	-0.6005	-0.7739
3.5	-0.0074	-0.0374	-0.0756	-0.1539	-0.3165	-0.4865	-0.6639	-0.8472
4.0	-0.0080	-0.0406	-0.0818	-0.1656	-0.3383	-0.5162	-0.7001	-0.8883

Table 5.25 Values of $[\log(f/p)]^{(2)}$ **(continued)**

T_r/p_r	1.2	1.5	2	3	5	7	10
0.55	8.8985	8.2742	6.9471	4.5090	-0.9354	-6.4940	-15.673
0.60	2.7077	1.6561	0.5013	-2.2661	-7.8239	-13.224	-22.499
0.65	-1.3275	-2.2139	-3.2442	-6.2818	-11.886	-17.527	-26.588
0.70	-3.9533	-4.4050	-5.8839	-8.3660	-13.990	-19.855	-28.772
0.75	-6.0960	-6.8688	-8.2569	-11.271	-17.505	-23.729	-33.192
0.80	-6.3813	-7.2268	-8.6649	-11.715	-18.009	-24.235	-33.586
0.85	-6.0906	-6.8611	-8.3583	-11.339	-17.380	-23.902	-33.228
0.90	-4.9387	-5.8284	-7.3646	-10.497	-16.652	-22.975	-32.321
0.93	-4.0990	-5.0369	-6.5885	-9.7268	-15.991	-22.225	-31.555
0.95	-3.4327	-4.3995	-6.0187	-9.1840	-15.509	-21.716	-31.042
0.97	-2.8014	-3.7844	-5.3722	-8.6177	-14.918	-21.198	-30.513
0.98	-2.4141	-3.3978	-5.0549	-8.2646	-14.661	-20.932	-30.227
0.99	-2.0574	-3.0813	-4.7226	-7.9808	-14.355	-20.641	-29.942
1.00	-1.6656	-2.682	-4.3679	-7.6489	-14.070	-20.362	-29.661
1.01	-1.2531	-2.3266	-4.0202	-7.3347	-13.794	-20.072	-29.361
1.02	-0.8708	-1.9331	-3.6774	-7.0161	-13.481	-19.773	-29.053
1.05	-0.0794	-0.8339	-2.6405	-6.0391	-12.577	-18.883	-28.180
1.10	0.6098	0.2320	-1.0271	-4.4542	-11.076	-17.417	-26.652
1.15	0.8898	0.6795	-0.1085	-3.0734	-9.6852	-15.990	-25.162
1.2	0.9772	0.8645	0.3571	-2.0330	-8.4069	-14.688	-23.764
1.3	0.9237	0.8724	0.5382	-1.0399	-6.4409	-12.434	-21.254
1.4	1.5114	1.6259	1.5815	0.6427	-3.5823	-9.0962	-17.650
1.5	1.1233	1.1780	1.0878	0.2775	-3.3023	-8.2295	-16.207
1.6	0.7823	0.7836	0.6381	-0.1708	-3.3633	-7.8127	-15.251
1.7	0.4938	0.4456	0.2393	-0.5950	-3.5783	-7.6551	-14.606
1.8	0.2456	0.1581	-0.0992	-0.9779	-3.8397	-7.6618	-14.203
1.9	0.0346	-0.0878	-0.3894	-1.3073	-4.1043	-7.7376	-13.950
2.0	-0.1418	-0.2955	-0.6350	-1.5918	-4.3511	-7.8472	-13.787
2.2	-0.4205	-0.6200	-1.0222	-2.0483	-4.7639	-8.0659	-13.606
2.4	-0.6220	-0.8555	-1.3015	-2.3757	-5.0675	-8.2394	-13.495
2.6	-0.7691	-1.0268	-1.5039	-2.6118	-5.2809	-8.3551	-13.385
2.8	-0.8770	-1.1515	-1.6502	-2.7772	-5.4240	-8.4162	-13.266
3.0	-0.9558	-1.2421	-1.7549	-2.8935	-5.5101	-8.4343	-13.130
3.5	-1.0366	-1.3306	-1.8497	-2.9708	-5.4775	-8.2174	-12.589
4.0	-1.0814	-1.3786	-1.8964	-2.9941	-5.4014	-8.0021	-12.110

Table 5.26 Values of $\left[(C_p - C_p^0)/R\right]^{(0)}$

T_r/p_r	0.01	0.05	0.1	0.2	0.4	0.6	0.8	1
0.55	0.0906	2.7103	2.7067	2.6986	2.6831	2.6681	2.6540	2.6407
0.60	0.0748	2.7923	2.7866	2.7751	2.7529	2.7322	2.7124	2.6926
0.65	0.0646	2.9330	2.9249	2.9078	2.8755	2.8448	2.8163	2.7891
0.70	0.0502	0.2634	3.1556	3.1112	3.0622	3.0164	2.9736	2.9335
0.75	0.0399	0.2202	0.4872	3.4110	3.3301	3.2569	3.1905	3.1289
0.80	0.0322	0.1755	0.3907	1.0046	3.7934	3.6630	3.5493	3.4482
0.85	0.0266	0.1427	0.3122	0.7650	4.6615	4.3753	4.1463	3.9569
0.90	0.0224	0.1183	0.2547	0.5997	1.8898	6.0016	5.3435	4.8863
0.93	0.0203	0.1066	0.2277	0.5256	1.5279	4.5998	6.9687	5.9514
0.95	0.0191	0.0998	0.2121	0.4840	1.3494	3.5059	9.5294	7.2177
0.97	0.0179	0.0937	0.1981	0.4474	1.2053	2.8360	10.7445	9.7673
0.98	0.0174	0.0908	0.1917	0.4307	1.1431	2.5898	7.9109	12.505
0.99	0.0169	0.0881	0.1856	0.4150	1.0858	2.3828	6.3326	19.348
1.00	0.0164	0.0855	0.1797	0.4003	1.0338	2.2057	5.3126	0.0000
1.01	0.0160	0.0830	0.1742	0.3863	0.9859	2.0523	4.5914	22.334
1.02	0.0156	0.0806	0.1689	0.3731	0.9415	1.9181	4.0507	13.161
1.05	0.0143	0.0741	0.1545	0.3376	0.8272	1.5992	3.0039	6.3522
1.10	0.0126	0.0649	0.1344	0.2893	0.6820	1.2417	2.0964	3.5309
1.15	0.0112	0.0573	0.118	0.2512	0.575	1.0051	1.5976	2.4462
1.2	0.0100	0.0510	0.1045	0.2204	0.4933	0.8375	1.2799	1.8581
1.3	0.0081	0.0411	0.0838	0.1742	0.3775	0.6162	0.8976	1.2300
1.4	0.0067	0.0337	0.0684	0.1408	0.2988	0.4762	0.6752	0.8978
1.5	0.0056	0.0282	0.0571	0.1166	0.2438	0.3823	0.5326	0.6950
1.6	0.0048	0.024	0.0483	0.0983	0.2032	0.3149	0.4335	0.5587
1.7	0.0041	0.0206	0.0414	0.0839	0.1722	0.2646	0.3611	0.4614
1.8	0.0036	0.0179	0.0359	0.0725	0.1478	0.2258	0.3062	0.3889
1.9	0.0031	0.0156	0.0314	0.0633	0.1284	0.1951	0.2634	0.3329
2.0	0.0028	0.0138	0.0277	0.0557	0.1125	0.1704	0.2292	0.2887
2.2	0.0022	0.0110	0.0220	0.0441	0.0886	0.1334	0.1785	0.2238
2.4	0.0018	0.0089	0.0178	0.0357	0.0715	0.1073	0.1432	0.1788
2.6	0.0015	0.0074	0.0147	0.0295	0.0589	0.0882	0.1174	0.1463
2.8	0.0012	0.0062	0.0124	0.0247	0.0493	0.0737	0.0980	0.1220
3.0	0.0011	0.0053	0.0105	0.0210	0.0419	0.0625	0.0830	0.1032
3.5	0.0007	0.0037	0.0073	0.0147	0.0291	0.0435	0.0576	0.0716
4.0	0.0005	0.0027	0.0054	0.0107	0.0214	0.0318	0.0422	0.0524

Table 5.26 Values of $\left[(C_p - C_p^0)/R\right]^{(0)}$ **(continued)**

T_r / p_r	1.2	1.5	2	3	5	7	10
0.55	2.6277	2.6090	2.5797	2.5284	2.4471	2.3870	2.3228
0.60	2.6741	2.6481	2.6077	2.5382	2.4308	2.3522	2.2682
0.65	2.7627	2.7265	2.6711	2.5775	2.4386	2.3393	2.2360
0.70	2.8962	2.8444	2.7676	2.6419	2.4622	2.3396	2.2159
0.75	3.0731	2.9970	2.8875	2.7157	2.4839	2.3344	2.1877
0.80	3.3584	3.2405	3.0780	2.8368	2.5358	2.3530	2.1811
0.85	3.7975	3.5985	3.3430	2.9951	2.6007	2.3773	2.1767
0.90	4.5446	4.1628	3.7244	3.2007	2.6778	2.4066	2.1747
0.93	5.3120	4.6821	4.0403	3.3536	2.7307	2.4258	2.1741
0.95	6.1047	5.1591	4.3019	3.4695	2.7685	2.4395	2.1737
0.97	7.3577	5.8047	4.6189	3.5988	2.8084	2.4538	2.1736
0.98	8.3145	6.2196	4.8024	3.6686	2.8288	2.4614	2.1738
0.99	9.6874	6.7212	5.0063	3.7419	2.8503	2.4687	2.1730
1.00	11.851	7.3395	5.2325	3.8185	2.8720	2.4763	2.1733
1.01	15.755	8.1169	5.4850	3.8998	2.8937	2.4843	2.1731
1.02	24.702	9.1165	5.7672	3.9844	2.9161	2.4915	2.1730
1.05	20.616	14.4718	6.8287	4.2622	2.9850	2.5143	2.1725
1.10	6.2447	14.5071	9.2631	4.7870	3.1027	2.5520	2.1717
1.15	3.7073	6.7724	9.7580	5.3010	3.2133	2.5855	2.1685
1.2	2.6214	4.2131	7.1288	5.5793	3.3025	2.6110	2.1626
1.3	1.6209	2.3230	3.6936	4.8413	3.3534	2.6231	2.1382
1.4	1.1448	1.5596	2.3320	3.5162	3.1892	2.5740	2.0996
1.5	0.8692	1.1509	1.6546	2.5469	2.8331	2.4386	2.0270
1.6	0.6901	0.8973	1.2581	1.9258	2.4244	2.2493	1.9309
1.7	0.5651	0.7259	1.0007	1.5156	2.0483	2.0348	1.8183
1.8	0.4734	0.6028	0.8212	1.2319	1.7331	1.8195	1.6974
1.9	0.4034	0.5106	0.6895	1.0262	1.4791	1.6191	1.5735
2.0	0.3487	0.4392	0.5892	0.8713	1.2750	1.4407	1.4519
2.2	0.2690	0.3366	0.4476	0.6559	0.9761	1.1499	1.2290
2.4	0.2144	0.2672	0.3534	0.5148	0.7731	0.9345	1.0415
2.6	0.1751	0.2177	0.2869	0.4165	0.6289	0.7732	0.8880
2.8	0.1458	0.1809	0.2379	0.3447	0.5226	0.6503	0.7639
3.0	0.1232	0.1528	0.2006	0.2903	0.4418	0.5546	0.6628
3.5	0.0854	0.1057	0.1385	0.2002	0.3068	0.3911	0.4815
4.0	0.0624	0.0773	0.1012	0.1464	0.2256	0.2905	0.3645

Table 5.27 Values of $\left[(C_p - C_p^0)/R\right]^{(1)}$

T_r / p_r	0.01	0.05	0.1	0.2	0.4	0.6	0.8	1
0.55	2.5167	14.429	14.433	14.445	14.462	14.480	14.495	14.510
0.60	1.2128	13.736	13.741	13.753	13.775	13.789	13.804	13.822
0.65	0.4351	13.083	13.088	13.100	13.127	13.154	13.169	13.185
0.70	0.2605	2.0365	7.9968	12.522	12.553	12.583	12.610	12.627
0.75	0.1657	0.8980	2.2060	13.033	13.086	13.125	13.152	13.175
0.80	0.1243	0.6258	1.3910	6.5856	12.563	12.641	12.689	12.720
0.85	0.0948	0.4738	0.9792	2.3804	12.124	12.348	12.463	12.521
0.90	0.0731	0.3664	0.7452	1.6228	5.2070	12.094	12.593	12.754
0.93	0.0628	0.3155	0.6401	1.3568	3.6693	11.601	12.893	13.353
0.95	0.0569	0.2860	0.5796	1.2163	3.0475	8.1819	12.853	14.293
0.97	0.0516	0.2596	0.5257	1.0950	2.6022	5.8278	27.235	16.474
0.98	0.0491	0.2473	0.5004	1.0403	2.4229	5.0829	16.992	18.826
0.99	0.0468	0.2358	0.4774	0.9895	2.2686	4.5014	11.699	24.880
1.00	0.0446	0.2248	0.4550	0.9411	2.1262	4.0436	8.9011	∞
1.01	0.0426	0.2144	0.4336	0.8958	1.9995	3.6692	7.1972	21.884
1.02	0.0406	0.2045	0.4137	0.8528	1.8847	3.3599	6.0631	12.502
1.05	0.0353	0.1778	0.3591	0.7375	1.5873	2.6697	4.1615	6.0253
1.10	0.0281	0.1415	0.2854	0.5834	1.2256	1.9512	2.7508	3.4809
1.15	0.0225	0.1133	0.2284	0.4646	0.9617	1.4902	2.0190	2.4689
1.2	0.0181	0.0913	0.1840	0.3724	0.7627	1.1613	1.5452	1.8685
1.3	0.0120	0.0602	0.1210	0.2435	0.4913	0.7346	0.9623	1.1570
1.4	0.0083	0.0417	0.0835	0.1669	0.3323	0.4902	0.6358	0.7613
1.5	0.0057	0.0286	0.0573	0.1142	0.2251	0.3302	0.4251	0.5084
1.6	0.0040	0.0199	0.0398	0.0788	0.1545	0.2252	0.2890	0.3445
1.7	0.0028	0.0140	0.0278	0.0549	0.1071	0.1549	0.1978	0.2352
1.8	0.0020	0.0098	0.0195	0.0384	0.0741	0.1068	0.1356	0.1604
1.9	0.0014	0.0069	0.0136	0.0266	0.0512	0.0728	0.0922	0.1089
2.0	0.0009	0.0047	0.0094	0.0182	0.0346	0.0492	0.0615	0.0716
2.2	0.0004	0.0020	0.0040	0.0077	0.0142	0.0193	0.0232	0.0263
2.4	0.0001	0.0006	0.0011	0.0020	0.0032	0.0036	0.0031	0.0021
2.6	0.0000	-0.0002	-0.0005	-0.0011	-0.0028	-0.0050	-0.0076	-0.0105
2.8	-0.0001	-0.0007	-0.0014	-0.0028	-0.0060	-0.0096	-0.0134	-0.0177
3.0	-0.0002	-0.0009	-0.0018	-0.0037	-0.0077	-0.0118	-0.0162	-0.0206
3.5	-0.0002	-0.0011	-0.0021	-0.0043	-0.0086	-0.0130	-0.0175	-0.0219
4.0	-0.0002	-0.0010	-0.0020	-0.0039	-0.0079	-0.0118	-0.0158	-0.0198

Table 5.27 Values of $\left[(C_p - C_p^0)/R\right]^{(1)}$ **(continued)**

T_r / p_r	1.2	1.5	2	3	5	7	10
0.55	14.522	14.539	14.566	14.600	14.634	14.628	14.579
0.60	13.836	13.852	13.876	13.904	13.916	13.897	13.844
0.65	13.200	13.218	13.237	13.251	13.234	13.194	13.117
0.70	12.647	12.660	12.675	12.668	12.616	12.540	12.424
0.75	13.187	13.193	13.188	13.137	12.995	12.853	12.684
0.80	12.734	12.736	12.704	12.595	12.343	12.131	11.902
0.85	12.537	12.521	12.432	12.201	11.787	11.488	11.189
0.90	12.774	12.689	12.456	11.986	11.320	10.908	10.530
0.93	13.355	13.131	12.675	11.941	11.069	10.583	10.159
0.95	14.186	13.706	12.957	11.954	10.911	10.372	9.9167
0.97	15.939	14.729	13.400	12.000	10.760	10.162	9.6785
0.98	17.550	15.531	13.710	12.040	10.687	10.060	9.5620
0.99	20.199	16.631	14.097	12.092	10.609	9.9570	9.4468
1.00	25.384	18.091	14.566	12.159	10.536	9.8535	9.3295
1.01	39.388	20.157	15.121	12.234	10.465	9.7464	9.2194
1.02	89.696	23.248	15.758	12.317	10.389	9.6470	9.1035
1.05	-2.7095	40.820	18.338	12.575	10.157	9.3350	8.7636
1.10	3.3123	-0.3776	20.573	12.868	9.7028	8.7909	8.2029
1.15	2.6090	1.4070	7.4414	12.123	9.1443	8.2160	7.6478
1.2	2.0470	1.7637	1.6655	9.3931	8.4298	7.6019	7.0897
1.3	1.2957	1.3570	1.1403	3.4611	6.4482	6.2980	5.9965
1.4	0.8579	0.9424	0.9230	1.3691	4.2034	4.8949	4.9306
1.5	0.5753	0.6457	0.6804	0.8159	2.6004	3.6689	4.0503
1.6	0.3907	0.4411	0.4827	0.5548	1.5914	2.6604	3.2931
1.7	0.2666	0.3020	0.3370	0.3854	0.9955	1.8945	2.6411
1.8	0.1813	0.2053	0.2290	0.2645	0.6412	1.3346	2.0880
1.9	0.1221	0.1371	0.1519	0.1730	0.4158	0.9385	1.6385
2.0	0.0798	0.0889	0.0965	0.1079	0.2671	0.6585	1.2825
2.2	0.0279	0.0286	0.0271	0.0237	0.0946	0.3161	0.7692
2.4	0.0002	-0.0028	-0.0098	-0.0214	0.0064	0.1364	0.4522
2.6	-0.0143	-0.0197	-0.0291	-0.0457	-0.0406	0.0374	0.2614
2.8	-0.0218	-0.0280	-0.0390	-0.0581	-0.0653	-0.0185	0.1401
3.0	-0.0251	-0.0321	-0.0433	-0.0629	-0.0778	-0.0498	0.0642
3.5	-0.0264	-0.0331	-0.0438	-0.0624	-0.0833	-0.0777	-0.0235
4.0	-0.0236	-0.0294	-0.0385	-0.0549	-0.0758	-0.0780	-0.0495

Table 5.28 Values of $\left[(C_p - C_p^0)/R\right]^{(2)}$

T_r / p_r	0.01	0.05	0.1	0.2	0.4	0.6	0.8	1
0.55	-126.08	-624.94	-625.807	-626.95	-628.96	-631.15	-633.10	-635.21
0.60	95.437	-577.04	-577.311	-578.22	-579.86	-580.97	-582.17	-583.62
0.65	48.850	-532.43	-532.583	-532.86	-533.93	-535.31	-535.88	-536.66
0.70	27.530	294.80	465.066	-492.15	-492.11	-492.71	-493.19	-492.99
0.75	17.254	128.29	554.974	-554.21	-553.66	-552.73	-551.55	-550.53
0.80	10.112	69.347	195.495	342.47	-510.68	-508.79	-506.62	-504.32
0.85	6.126	39.599	104.018	347.98	-482.36	-479.56	-474.88	-470.29
0.90	3.837	23.603	59.067	174.67	865.85	-471.05	-462.90	-452.03
0.93	2.943	17.647	43.027	121.71	472.36	3170.8	-475.86	-454.77
0.95	2.479	14.633	35.165	96.779	345.01	1159.4	-517.72	-475.45
0.97	2.097	12.214	28.913	77.635	260.16	695.54	5772.4	-540.20
0.98	1.935	11.190	26.325	69.812	227.52	569.82	1959.5	-593.16
0.99	1.785	10.243	23.919	62.793	199.89	477.11	1229.5	-736.28
1.00	1.648	9.410	21.822	56.651	176.43	404.75	905.09	∞
1.01	1.525	8.646	19.963	51.216	156.21	347.23	713.47	1483.0
1.02	1.414	7.945	18.237	46.361	138.63	300.09	582.03	1058.6
1.05	1.132	6.258	14.106	34.832	99.248	200.88	353.16	560.15
1.10	0.794	4.304	9.468	22.298	58.835	111.16	180.73	266.21
1.15	0.577	3.057	6.556	14.849	36.704	65.662	101.59	144.06
1.2	0.430	2.231	4.675	10.258	23.926	40.903	60.726	83.620
1.3	0.256	1.294	2.628	5.436	11.395	17.877	24.775	32.302
1.4	0.041	0.208	0.432	0.954	2.114	3.658	5.564	7.974
1.5	-0.003	-0.017	-0.038	-0.087	-0.144	-0.172	0.032	0.437
1.6	-0.028	-0.143	-0.289	-0.583	-1.227	-1.863	-2.383	-2.754
1.7	-0.042	-0.214	-0.429	-0.873	-1.797	-2.658	-3.481	-4.152
1.8	-0.051	-0.251	-0.505	-1.020	-2.033	-3.046	-3.964	-4.762
1.9	-0.055	-0.274	-0.546	-1.086	-2.163	-3.168	-4.145	-5.024
2.0	-0.056	-0.281	-0.563	-1.114	-2.187	-3.229	-4.167	-5.015
2.2	-0.056	-0.280	-0.558	-1.096	-2.139	-3.097	-4.014	-4.872
2.4	-0.054	-0.268	-0.532	-1.047	-2.027	-2.945	-3.787	-4.568
2.6	-0.051	-0.253	-0.499	-0.986	-1.906	-2.759	-3.561	-4.315
2.8	-0.048	-0.237	-0.468	-0.923	-1.785	-2.588	-3.336	-4.011
3.0	-0.045	-0.222	-0.440	-0.865	-1.672	-2.432	-3.135	-3.795
3.5	-0.038	-0.190	-0.377	-0.741	-1.441	-2.099	-2.721	-3.307
4.0	-0.033	-0.165	-0.327	-0.648	-1.260	-1.846	-2.399	-2.923

Table 5.28 Values of $\left[(C_p - C_p^0)/R\right]^{(2)}$ **(continued)**

T_r/p_r	1.2	1.5	2	3	5	7	10
0.55	-637.06	-639.71	-644.12	-651.45	-663.48	-671.89	-679.95
0.60	-585.05	-586.64	-589.52	-594.35	-601.84	-607.74	-614.66
0.65	-537.23	-538.31	-539.48	-541.56	-545.24	-548.53	-553.28
0.70	-493.62	-493.41	-493.50	-492.41	-492.39	-493.11	-495.44
0.75	-549.54	-547.61	-544.71	-539.34	-532.48	-528.73	-527.96
0.80	-501.69	-498.05	-492.03	-482.13	-468.89	-462.00	-459.05
0.85	-465.37	-457.95	-446.77	-429.66	-409.74	-400.38	-395.61
0.90	-441.24	-426.47	-406.58	-380.06	-353.13	-342.17	-336.97
0.93	-434.82	-410.47	-381.64	-347.97	-318.82	-308.11	-303.76
0.95	-437.59	-399.92	-362.63	-324.65	-295.13	-285.38	-281.93
0.97	-456.18	-391.25	-339.92	-298.65	-270.78	-262.28	-260.60
0.98	-472.62	-387.14	-326.80	-284.11	-258.27	-251.10	-250.38
0.99	-477.96	-379.99	-311.98	-268.65	-245.02	-239.37	-239.70
1.00	-458.40	-355.06	-293.52	-252.44	-231.85	-227.83	-229.23
1.01	-348.10	-306.05	-268.18	-234.38	-218.23	-215.73	-219.50
1.02	2340.8	-211.90	-230.75	-214.38	-204.04	-204.09	-208.97
1.05	825.29	1357.77	-21.610	-136.04	-158.50	-167.84	-178.56
1.10	367.57	571.13	732.56	66.169	-70.591	-103.97	-128.79
1.15	194.16	299.62	496.54	306.46	26.831	-37.928	-80.274
1.2	109.83	162.96	302.71	425.61	120.38	26.997	-33.076
1.3	40.762	57.195	102.96	243.17	228.31	127.62	49.950
1.4	11.305	18.402	39.764	123.54	246.65	200.72	132.55
1.5	1.420	3.558	12.466	51.530	157.28	176.07	146.08
1.6	-2.859	-2.202	1.073	20.329	89.730	130.37	130.71
1.7	-4.710	-4.982	-4.120	6.096	49.580	87.798	104.77
1.8	-5.433	-6.144	-6.209	-1.290	26.051	56.496	78.613
1.9	-5.752	-6.604	-7.237	-4.815	12.373	34.954	56.361
2.0	-5.781	-6.741	-7.658	-6.905	4.138	20.348	38.205
2.2	-5.601	-6.538	-7.742	-8.514	-4.205	4.308	15.391
2.4	-5.248	-6.233	-7.483	-8.899	-7.871	-3.770	2.617
2.6	-4.947	-5.876	-7.162	-8.836	-9.496	-7.868	-4.755
2.8	-4.657	-5.567	-6.812	-8.623	-10.255	-10.114	-9.027
3.0	-4.414	-5.257	-6.506	-8.381	-10.556	-11.387	-11.694
3.5	-3.863	-4.627	-5.789	-7.719	-10.585	-12.643	-14.994
4.0	-3.432	-4.143	-5.237	-7.127	-10.164	-12.732	-16.022

5.8.2 Comparisons and Results

Comparisons between calculated and the highly accurate equations of state for the thermodynamic properties indicate that highly accurate results for the polar fluids are obtained as well as for the simple and nonpolar substances. Comparison of the calculated values for the compressibility factor, enthalpy, entropy, fugacity, heat capacity for argon, methane, nitrogen, ethane, chlorodifluoromethane (R22), ammonia, 1,1,1-trifluoroethane (R143a), 1,1-difluoroethane (R152a), difluoromethane (R32), and water with recommended data are shown in Figs. 5.9 to 5.60, respectively. These results indicate that the extended theory may well yield a complete description of all classes of molecules at least one order of magnitude more accurate than those obtained from the corresponding-states theory of Pitzer et al.

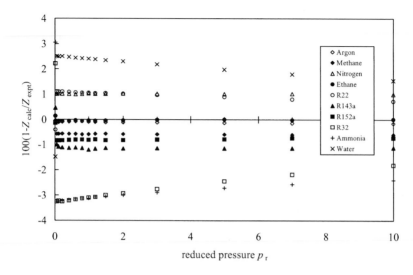

Fig. 5.9 Percentage deviations of the recommended compressibility-factor data with values calculated from the extended corresponding-states theory along the isotherm of the reduced temperature of 0.55 for some representative molecules.

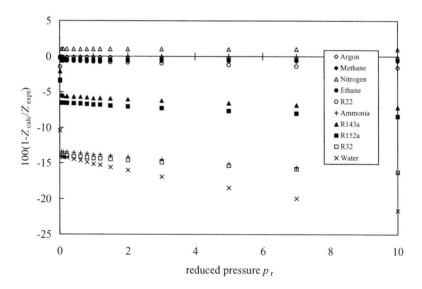

Fig. 5.10 Percentage deviations of the recommended compressibility-factor data with values calculated from the corresponding-states theory of Pitzer et al. along the isotherm of the reduced temperature of 0.55 for some representative molecules.

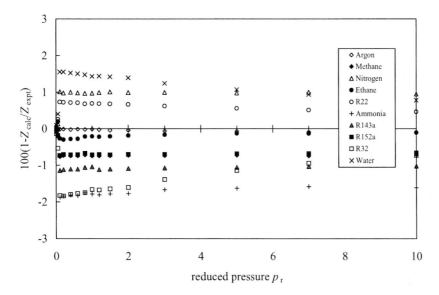

Fig. 5.11 Percentage deviations of the recommended compressibility-factor data with values calculated from the extended corresponding-states theory along the isotherm of the reduced temperature of 0.7 for some representative molecules.

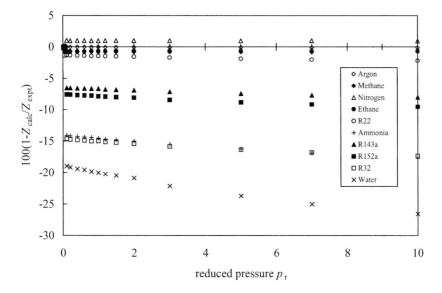

Fig. 5.12 Percentage deviations of the recommended compressibility-factor data with values calculated from the corresponding-states theory of Pitzer et al. along the isotherm of the reduced temperature of 0.55 for some representative molecules.

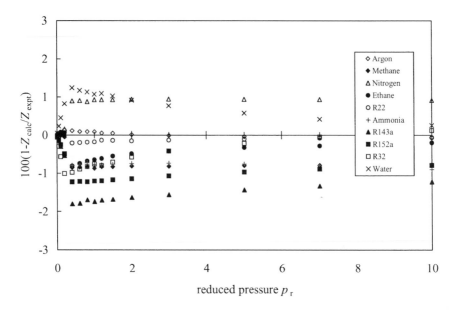

Fig. 5.13 Percentage deviations of the recommended compressibility-factor data with values calculated from the extended corresponding-states theory along the isotherm of the reduced temperature of 0.85 for some representative molecules.

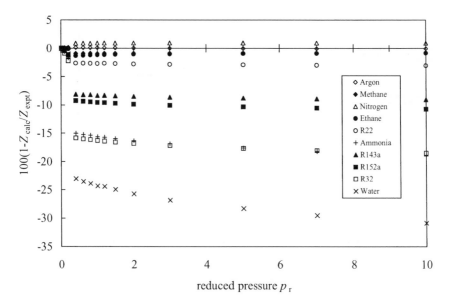

Fig. 5.14 Percentage deviations of the recommended compressibility-factor data with values calculated from the corresponding-states theory of Pitzer et al. along the isotherm of the reduced temperature of 0.85 for some representative molecules.

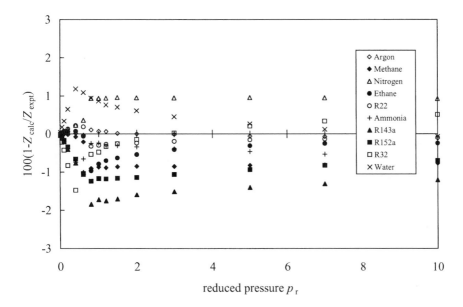

Fig. 5.15 Percentage deviations of the recommended compressibility-factor data with values calculated from the extended corresponding-states theory along the isotherm of the reduced temperature of 0.93 for some representative molecules.

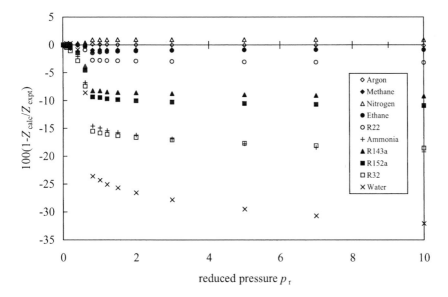

Fig. 5.16 Percentage deviations of the recommended compressibility-factor data with values calculated from the corresponding-states theory of Pitzer et al. along the isotherm of the reduced temperature of 0.93 for some representative molecules.

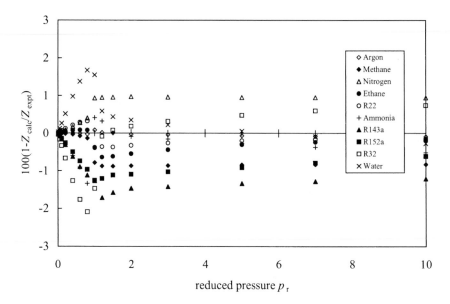

Fig. 5.17 Percentage deviations of the recommended compressibility-factor data with values calculated from the extended corresponding-states theory along the critical isotherm for some representative molecules.

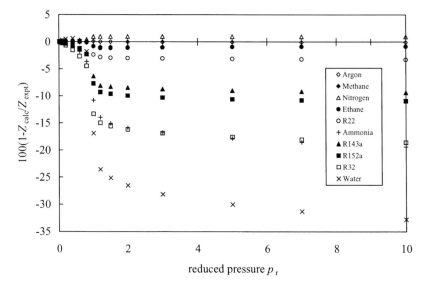

Fig. 5.18 Percentage deviations of the recommended compressibility-factor data with values calculated from the corresponding-states theory of Pitzer et al. along the critical isotherm for some representative molecules.

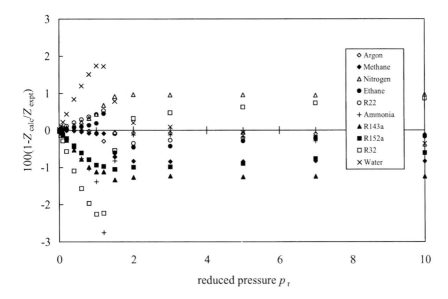

Fig. 5.19 Percentage deviations of the recommended compressibility-factor data with values calculated from the extended corresponding-states theory along the isotherm of the reduced temperature of 1.05 for some representative molecules.

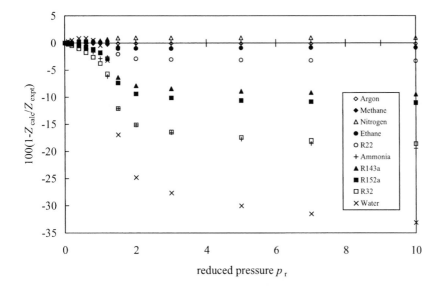

Fig. 5.20 Percentage deviations of the recommended compressibility-factor data with values calculated from the corresponding-states theory of Pitzer et al. along the isotherm of the reduced temperature of 1.05 for some representative molecules.

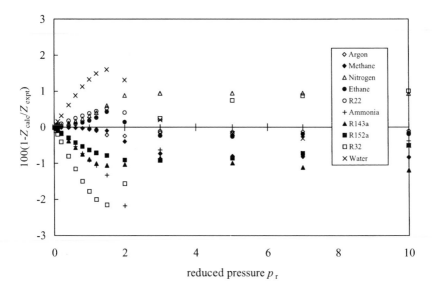

Fig. 5.21 Percentage deviations of the recommended compressibility-factor data with values calculated from the extended corresponding-states theory along the isotherm of the reduced temperature of 1.15 for some representative molecules.

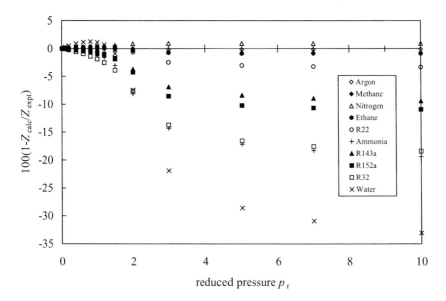

Fig. 5.22 Percentage deviations of the recommended compressibility-factor data with values calculated from the corresponding-states theory of Pitzer et al. along the isotherm of the reduced temperature of 1.15 for some representative molecules.

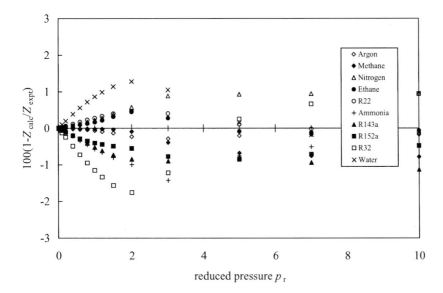

Fig. 5.23 Percentage deviations of the recommended compressibility-factor data with values calculated from the extended corresponding-states theory along the isotherm of the reduced temperature of 1.30 for some representative molecules.

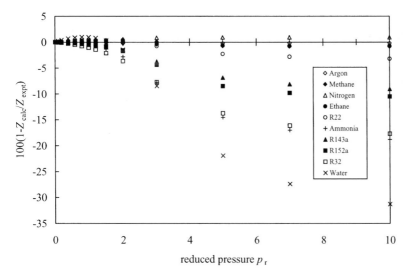

Fig. 5.24 Percentage deviations of the recommended compressibility-factor data with values calculated from the corresponding-states theory of Pitzer et al. along the isotherm of the reduced temperature of 1.30 for some representative molecules.

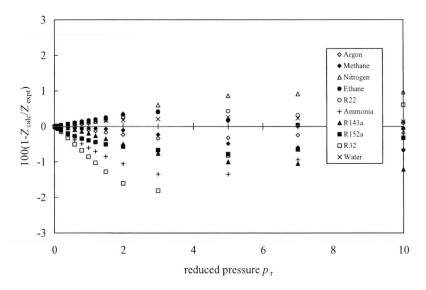

Fig. 5.25 Percentage deviations of the recommended compressibility-factor data with values calculated from the extended corresponding-states theory along the isotherm of the reduced temperature of 1.50 for some representative molecules.

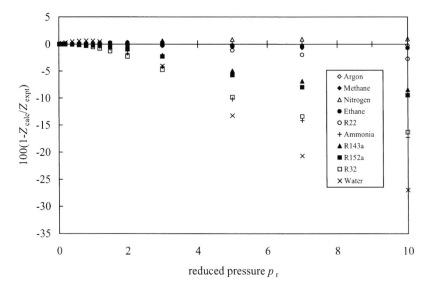

Fig. 5.26 Percentage deviations of the recommended compressibility-factor data with values calculated from the corresponding-states theory of Pitzer et al. along the isotherm of the reduced temperature of 1.50 for some representative molecules.

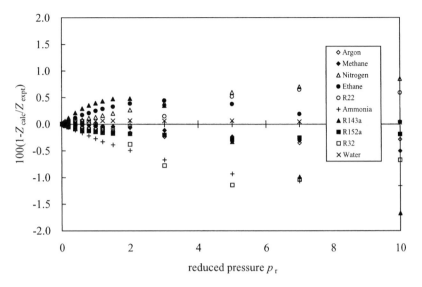

Fig. 5.27 Percentage deviations of the recommended compressibility-factor data with values calculated from the extended corresponding-states theory along the isotherm of the reduced temperature of 2 for some representative molecules.

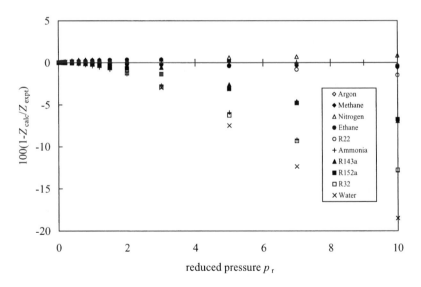

Fig. 5.28 Percentage deviations of the recommended compressibility-factor data with values calculated from the corresponding-states theory of Pitzer et al. along the isotherm of the reduced temperature of 2 for some representative molecules.

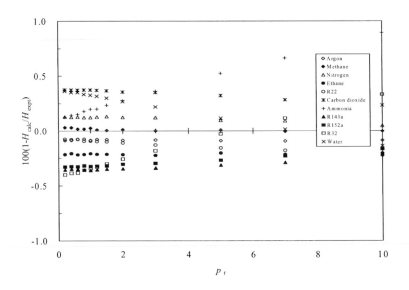

Fig. 5.29 Percentage deviations of the recommended enthalpy-departure data with values calculated from the extended corresponding-states theory along the isotherm of the reduced temperature of 0.75 for some representative molecules.

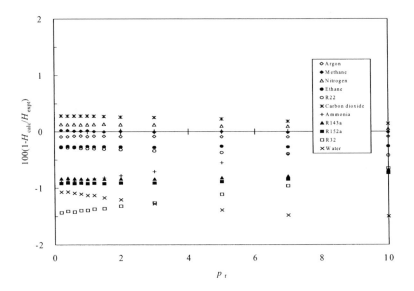

Fig. 5.30 Percentage deviations of the recommended enthalpy-departure data with values calculated from the corresponding-states theory of Pitzer et al. along the isotherm of the reduced temperature of 0.75 for some representative molecules.

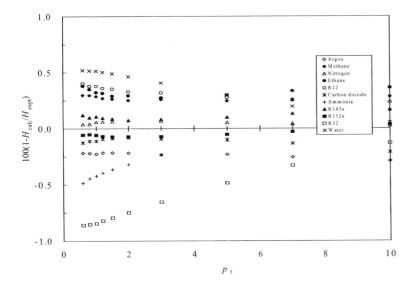

Fig. 5.31 Percentage deviations of the recommended enthalpy-departure data with values calculated from the extended corresponding-states theory along the isotherm of the reduced temperature of 0.9 for some representative molecules.

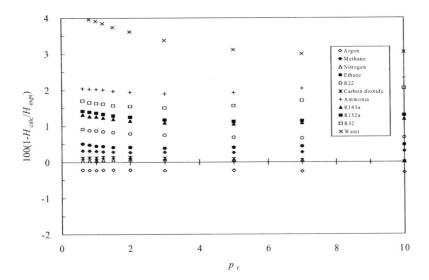

Fig. 5.32 Percentage deviations of the recommended enthalpy-departure data with values calculated from the corresponding-states theory of Pitzer et al. along the isotherm of the reduced temperature of 0.9 for some representative molecules.

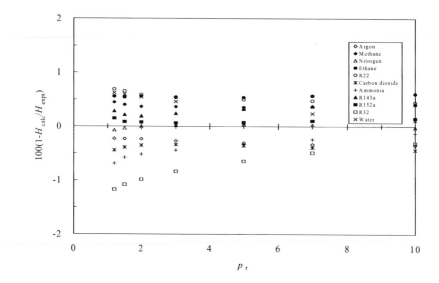

Fig. 5.33 Percentage deviations of the recommended enthalpy-departure data with values calculated from the extended corresponding-states theory along the critical isotherm for some representative molecules.

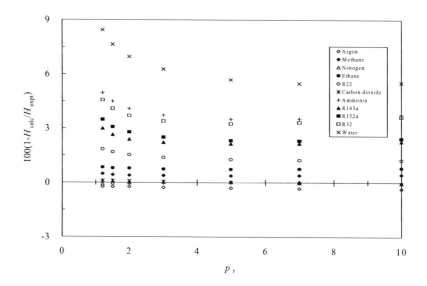

Fig. 5.34 Percentage deviations of the recommended enthalpy-departure data with values calculated from the corresponding-states theory of Pitzer et al. along the critical isotherm for some representative molecules

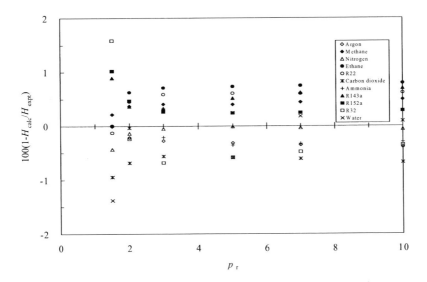

Fig. 5.35 Percentage deviations of the recommended enthalpy-departure data with values calculated from the extended corresponding-states theory along the isotherm of the reduced temperature of 1.1 for some representative molecules.

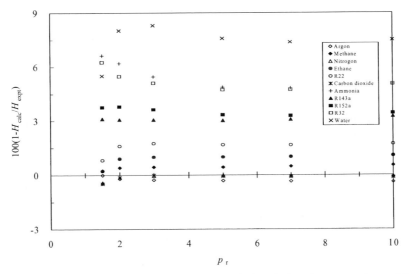

Fig. 5.36 Percentage deviations of the recommended enthalpy-departure data with values calculated from the corresponding-states theory of Pitzer et al. along the isotherm of the reduced temperature of 1.1 for some representative molecules.

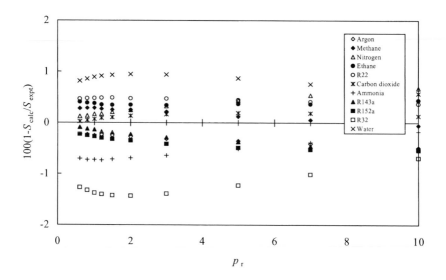

Fig. 5.37 Percentage deviations of the recommended entropy-departure data with values calculated from the extended corresponding-states theory along the isotherm of the reduced temperature of 0.75 for some representative molecules.

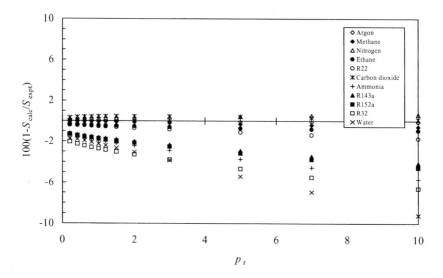

Fig. 5.38 Percentage deviations of the recommended entropy-departure data with values calculated from the corresponding-states theory of Pitzer et al. along the isotherm of the reduced temperature of 0.75 for some representative molecules.

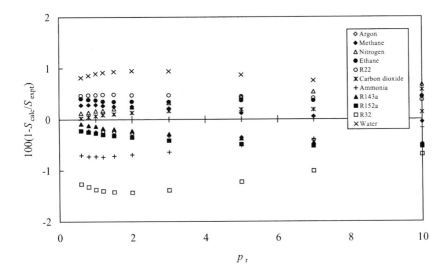

Fig. 5.39 Percentage deviations of the recommended entropy-departure data with values calculated from the extended corresponding-states theory along the isotherm of the reduced temperature of 0.9 for some representative molecules.

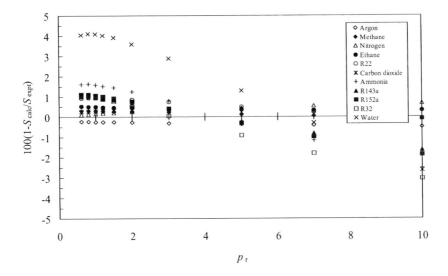

Fig. 5.40 Percentage deviations of the recommended entropy-departure data with values calculated from the corresponding-states theory of Pitzer et al. along the isotherm of the reduced temperature of 0.9 for some representative molecules.

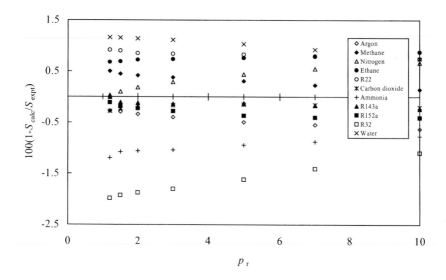

Fig. 5.41 Percentage deviations of the recommended entropy-departure data with values calculated from the extended corresponding-states theory along the critical isotherm for some representative molecules.

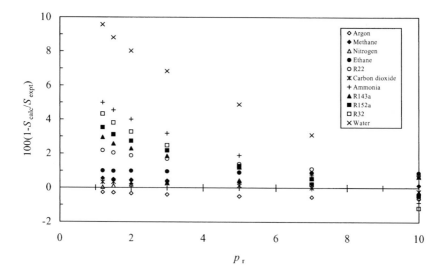

Fig. 5.42 Percentage deviations of the recommended entropy-departure data with values calculated from the corresponding-states theory of Pitzer et al. along the critical isotherm for some representative molecules

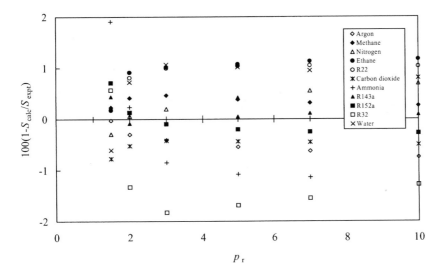

Fig. 5.43 Percentage deviations of the recommended entropy-departure data with values calculated from the extended corresponding-states theory along the isotherm of the reduced temperature of 1.1 for some representative molecules.

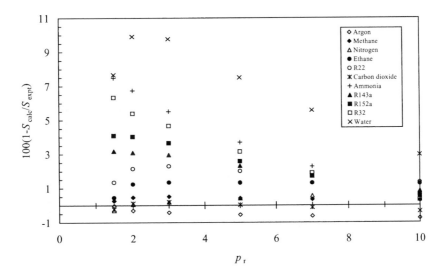

Fig. 5.44 Percentage deviations of the recommended entropy-departure data with values calculated from the corresponding-states theory of Pitzer et al. along the isotherm of the reduced temperature of 1.1 for some representative molecules.

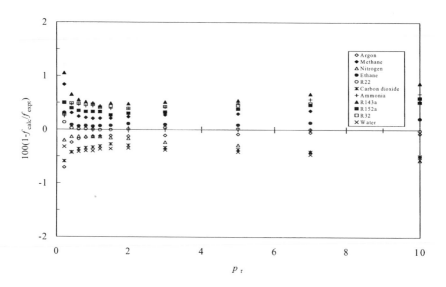

Fig. 5.45 Percentage deviations of the recommended fugacity-coefficient data with values calculated from the extended corresponding-states theory along the isotherm of the reduced temperature of 0.75 for some representative molecules.

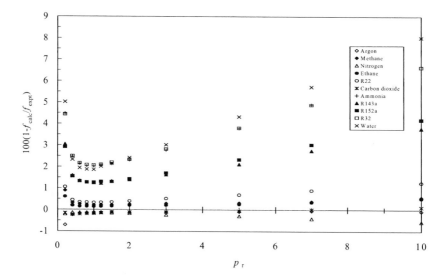

Fig. 5.46 Percentage deviations of the recommended fugacity-coefficient data with values calculated from the corresponding-states theory of Pitzer et al. along the isotherm of the reduced temperature of 0.75 for some representative molecules

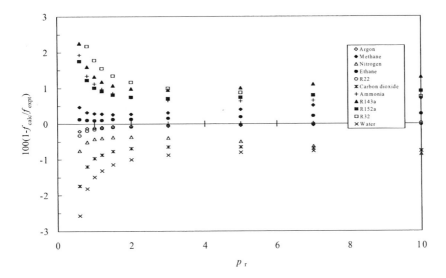

Fig. 5.47 Percentage deviations of the recommended fugacity-coefficient data with values calculated from the extended corresponding-states theory along the isotherm of the reduced temperature of 0.9 for some representative molecules.

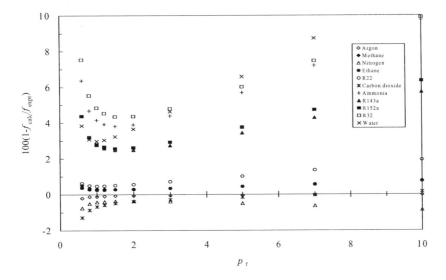

Fig. 5.48 Percentage deviations of the recommended fugacity-coefficient data with values calculated from the corresponding-states theory of Pitzer et al. along the isotherm of the reduced temperature of 0.9 for some representative molecules

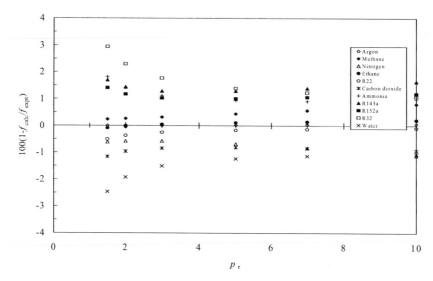

Fig. 5.49 Percentage deviations of the recommended fugacity-coefficient data with values calculated from the extended corresponding-states theory along the critical isotherm for some representative molecules.

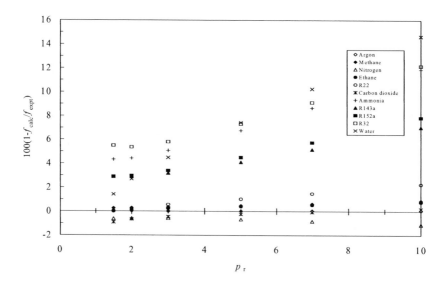

Fig. 5.50 Percentage deviations of the recommended fugacity-coefficient data with values calculated from the corresponding-states theory of Pitzer et al. along the critical isotherm for some representative molecules

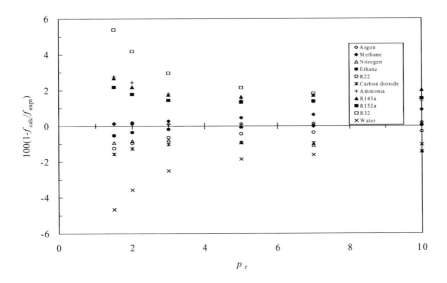

Fig. 5.51 Percentage deviations of the recommended fugacity-coefficient data with values calculated from the extended corresponding-states theory along the isotherm of the reduced temperature of 1.1 for some representative molecules.

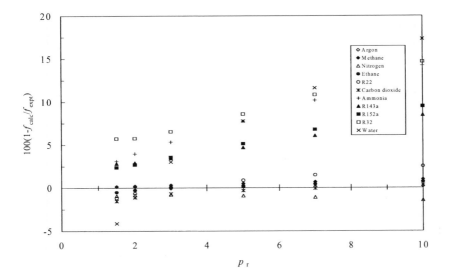

Fig. 5.52 Percentage deviations of the recommended fugacity-coefficient data with values calculated from the corresponding-states theory of Pitzer et al. along the isotherm of the reduced temperature of 1.1 for some representative molecules

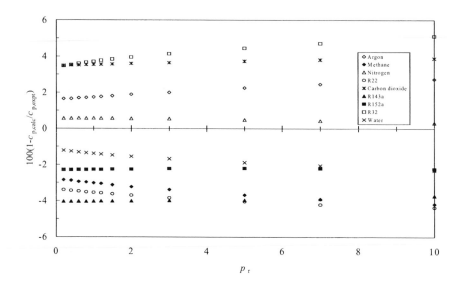

Fig. 5.53 Percentage deviations of the isobaric-heat-capacity data with values calculated from the extended corresponding-states theory along the isotherm of the reduced temperature of 0.75 for some representative molecules.

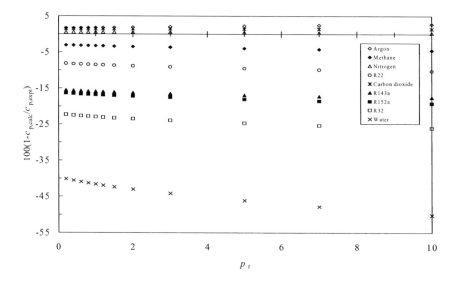

Fig. 5.54 Percentage deviations of the recommended isobaric-heat-capacity data with values calculated from the corresponding-states theory of Pitzer et al. along the isotherm of the reduced temperature of 0.75 for some representative molecules.

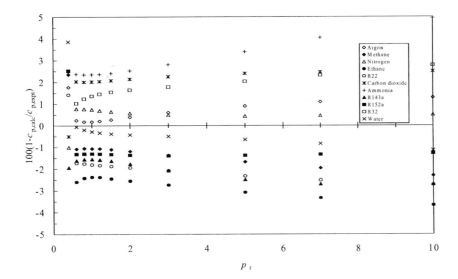

Fig. 5.55 Percentage deviations of the isobaric-heat-capacity data with values calculated from the extended corresponding-states theory along the isotherm of the reduced temperature of 0.9 for some representative molecules.

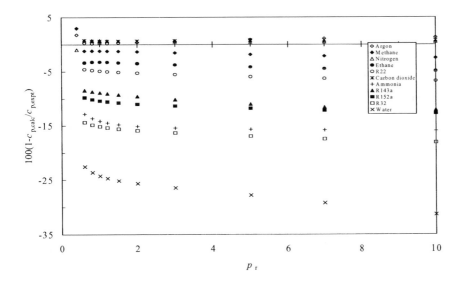

Fig. 5.56 Percentage deviations of the recommended isobaric-heat-capacity data with values calculated from the corresponding-states theory of Pitzer et al. along the isotherm of the reduced temperature of 0.9 for some representative molecules.

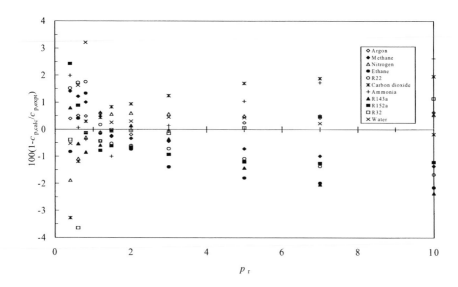

Fig. 5.57 Percentage deviations of the isobaric-heat-capacity data with values calculated from the extended corresponding-states theory along the critical isotherm for some representative molecules.

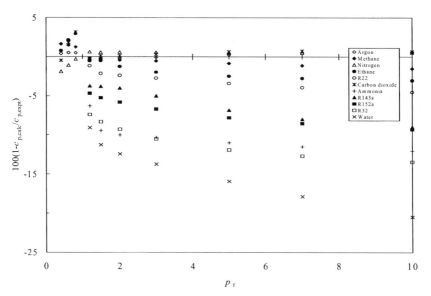

Fig. 5.58 Percentage deviations of the recommended isobaric-heat-capacity data with values calculated from the corresponding-states theory of Pitzer et al. along the critical isotherm for some representative molecules.

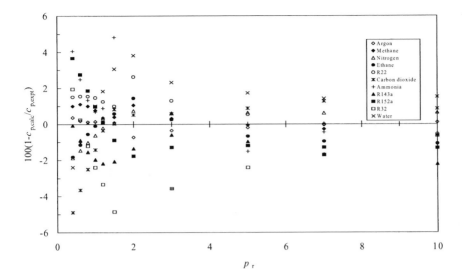

Fig. 5.59 Percentage deviations of the isobaric-heat-capacity data with values calculated from the extended corresponding-states theory along the isotherm of the reduced temperature of 1.1 for some representative molecules.

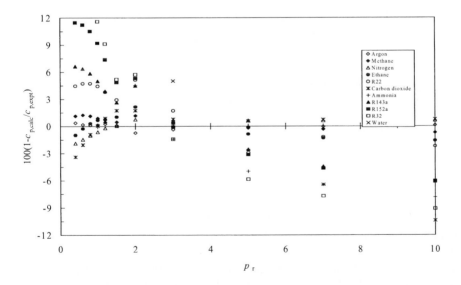

Fig. 5.60 Percentage deviations of the recommended isobaric-heat-capacity data with values calculated from the corresponding-states theory of Pitzer et al. along the isotherm of the reduced temperature of 1.1 for some representative molecules.

References

Abusleme, J. A. and J. H. Vera, 1989, *Fluid Phase Equil.* **45**: 287.

Adachi, Y. and B. C. -Y. Lu, 1984, *AIChE J.* **30**: 991.

Adachi, Y., B. C. -Y. Lu, and H. Sugie, 1983, *Fluid Phase Equil.* **11**: 29.

Albright, P. C., J. V. Sengers, J. F. Nicoll, and M. Ley-Koo, 1986, *Int. J. Thermophys.* **7**: 75.

Ambrose, D. and C. Tsonopoulos, 1995, *J. Chem. Eng. Data* **40**: 531.

Anisimov, M. A., S. B. Kiselev, J. V. Sengers, and S. Tang, 1992, *Physica A* **188**: 487.

Baehr, H. D. and R. Tillner-Roth, 1995, Thermodynamic Properties of Environmentally Acceptable Refrigerants (Springer, Berlin).

Bagnuls, C. and C. Bervillier, 1985, *Phys. Rev. B* **32**: 7209.

Bagnuls, C., C. Bervillier, D. I. Meiron, and B. G. Nickel, 1987, *Phys. Rev. B* **35**: 3585.

Barker, J. A. and D. Henderson, 1976, *Rev. Mod. Phys.* **48**: 587.

Beattie, J. A. and O. C. Bridgemann, 1928, *Proc. Am. Acad. Arts Sci.* **63**: 229.

Benedict, M., G. B. Webb, and L. C. Rubin, 1940, *J. Chem. Phys.* **8**: 334.

Bhirud, V. L., 1978, *AIChE J.* **24**: 880, 1127.

Bignell, C. M. and P. J. Dunlop, 1993, *J. Chem. Eng. Data* **38**: 139.

Bignell, C. M. and P. J. Dunlop, 1993, *J. Chem. Phys.* **98**: 4889.

Blanke, W. and R. Weiss, 1992, *Fluid Phase Equil.* **80**: 179.

Boyes, S. J. and L. A. Weber, 1995, *J. Chem. Thermodyn.* **27**: 163.

Boyes, S. J., M. B. Ewing, and A. R. H. Goodwin, 1992, *J. Chem. Thermodyn.* **24**: 1151.

Burnside, B. M, 1971, *Chem. Ind.* **40**: 1108.

Campbell, A. N. and R. M. Chatterjee, 1968, *Can. J. Chem.* **46**: 575.

Campbell, S. W. and G. Thodos, 1985, *J. Chem. Eng. Data* **30**: 102.

Carnahan, N. F. and K. E. Starling, 1972, *AIChE J.* **18**: 1184.

Carnahan, N. F. and K. E. Starling, 1969, *J. Chem. Phys.* **51**: 635.

Chen, Z. Y., A. Abbaci, S. Tang, and J. V. Sengers, 1990, *Phys. Rev. A* **42**: 4470.

Chen, Z. Y., P. C. Albright, and J. V. Sengers, 1990, *Phys. Rev. A* **41**: 3161.

Cragoe, C. S., E. C. McKelvey, and G. F. O'Connor, 1922, *Sci. Pap. Bur. Stands.* **18**: 707.

Defibaugh, D. R. and G. Morrison, 1992, *Fluid Phase Equil.* **80**: 157.

Defibaugh, D. R., G. Morrison, and L. A. Weber, 1994, *J. Chem. Eng. Data* **39**: 333.

Demiriz, A. M., R. Kohlen, C. Koopmann, D. Moeller, P. Sauermann, G. A. Iglesias-Silva, and F. Kohler, 1993, *Fluid Phase Equil.* **85**: 313.

De Santis, R., F. Gironi, and L. Marrelli, 1976, *Ind. Eng. Chem. Fundam.* **15**: 183.

Dohm, V., 1987, *J. Low Temp. Phys.* **69**: 51.

Duschek, W., R. Kleinrahm, and W. Wagner, 1990, *J. Chem. Thermodyn.* **22**: 841.

Dymond, J. H., 1986, *Fluid Phase Equil.* **27**: 1.

Dymond, J. H. and E. B. Smith, 1980, The Virial Coefficients of Pure Gases and Mixtures, A Critical Compilation (Clarendon Press, Oxford).

Easteal, A. J. and L. A. Woolf, 1982, *J. Chem. Thermodyn.* **14**: 755.

Edison, T. A. and J. V. Sengers, 1999, *Int. J. Refrig.* **22**: 365.

Eubank, P. T., L. L. Joffrion, M. R. Patel, and W. Warowny, 1988, *J. Chem. Thermodyn.* **20**: 1009.

Friend, D. G., H. Ingham, and J. F. Ely, 1991, *J. Phys. Chem. Ref. Data* **20**: 275.

Fukuizumi, H. and M. Uematsu, 1992, *Int. J. Thermophys.* **12**: 371.

Fuller, G. G., 1976, *Ind. Eng. Chem. Fundam.* **15**: 254.

Gilgen, R., R. Kleinrahm, and W. Wagner, 1994, *J. Chem. Thermodyn.* **26**: 399.

Goodwin, A. R. H. and M. R. Moldover, 1990, *J. Chem. Phys.* **93**: 2741.

Goodwin, A. R. H. and M. R. Moldover, 1991, *J. Chem. Phys.* **95**: 5236.

Goodwin, R. D., 1987, *J. Phys. Chem. Ref. Data* **16**: 799.

Goodwin, R. D., 1988, *J. Phys. Chem. Ref. Data* **17**: 1541.

Gude, M. and A. S. Teja, 1995, *J. Chem. Eng. Data* **40**: 1025.

Gunn, R. D. and T. Yamada, 1971, *AIChE J.* **17**: 1341.

Guo, T. M. and L. Du, 1989, *Fluid Phase Equil.* **52**: 47.

Guo, T. M., C. H. Kim, and K. C. Chao, 1985, *Ind. Eng. Chem. Process Des. Dev.* **24**: 764.

Haar, L. and J. S. Gallagher, 1978, *J. Phys. Chem. Ref. Data* **7**: 635.

Haendel, G., R. Kleinrahm, and W. Wagner, 1992, *J. Chem. Thermophys.* **24**: 697.

Hales, J. L. and J. H. Ellender, 1976, *J. Chem. Thermodyn.* **8**: 1177.

Hankinson, R. W. and G. H. Thomson, 1979, *AIChE J.* **25**: 653.

Harmens, A. and H. Knapp, 1980, *Ind. Eng. Chem. Fundam.* **19**: 291.

Hayden, J. G. and J. P. O'Connell, 1975, *Ind. Eng. Chem. Process Des. Dev.* **14**: 209.

Haynes, W. M. and M. J. Hiza, 1977, *J. Chem. Thermodyn.* **9**: 179.

Helfand, E., H. L. Frisch, and J. L. Lebowitz, 1961, *J. Chem. Phys.* **34**: 1037.

Henderson, D., 1979, Equation of State in Engineering and Research, 176th ACS Meeting, Washington D.C.

Higashi, Y., 1994, *Int. J. Refrig.* **17**: 524.

Hirschfelder J. O., C. F. Curtiss, and R. B. Bird, 1964, Molecular Theory of Gases and Liquids (Wiley, New York).

Holcomb, C. D., V. G. Niesen, L. J. Van Poolen, and S. L. Outcalt, 1993, *Fluid Phase Equil.* **91**: 145.

Hou, Y. C., B. Zhang, and H. Q. Tang, 1981, *J. Chem. Ind. Eng. China* **1**: 1.

Hsu, C. C. and J. J. McKetta, 1964, *J. Chem. Eng. Data* **6**: 45.

IAPWS, 1994, Skeleton Tables 1985 for the Thermodynamic Properties of Ordinary Water Substance.

Ishikawa, T., W. K. Chung, and B. C. -Y. Lu, 1980, *AIChE J.* **26**: 312.

Iwai, Y., M. R. Margerum, and B. C. -Y. Lu, 1988, *Fluid Phase Equil.* **42**: 21.

Jacobsen, R. T. and R. J. Stewart, 1973, *J. Phys. Chem. Ref. Data* **2**: 757.

Jin, G. X., S. Tang, and J. V. Sengers, 1992, *Int. J. Thermophys.* **13**: 671.

Kay, W. B. and W. E. Donham, 1955, *Chem. Eng. Sci.* **4**: 1.

Kasahara, K., T. Munakata, and M. Uematsu, 1999, *J. Chem. Thermodyn.* **31**: 1273.

Kerl, K. and H. Varchmin, 1991, *Int. J. Thermophys.* **12**: 171.

Kiselev, S. B. and J. V. Sengers, 1993, *Int. J. Thermophys.* **14**: 1.

Kiselev, S. B., 1988, *High Temp.* **28**: 42.

Kiselev, S. B., I. G. Kostyukova, and A. A. Povodyrev, 1991, *Int. J. Thermophys.* **12**: 877.

Kleinrahm, R. and W. Wagner, 1986, *J. Chem. Thermodyn.* **18**: 739.

Kobe, K. A. and R. E. Lynn, 1953, *Chem. Rev.* **52**: 117.

Kobe, K. A., H. R. Crawford, and R. W. Stephenson, 1955, *Ind. Eng. Chem.* **47**: 1767.

Kohlen, R., H. Kratzke, and S. Muller, 1985, *J. Chem. Thermodyn.* **17**: 1141.

Kratzke, H. and S. Mueller, 1984, *J. Chem. Thermodyn.* **16**: 1157.

Kratzke, H. and S. Mueller, 1985, *J. Chem. Thermodyn.* **17**: 151.

Kreglewski, A., 1969, *J. Phys. Chem.* **73**: 608.

Kubic, W. L., 1982, *Fluid Phase Equil.* **9**: 79.

Kubic, W. L., 1986, *Fluid Phase Equil.* **31**: 35.

Kudchadker, A. P., G. H. Alani, and B. J. Zwolinski, 1968, *Chem. Rev.* **68**: 659.

Leach, J. W., P. S. Chappelear, and T. W. Leland, 1968, *AIChE J.* **14**: 568.

Lee, B. I. and M. G. Kesler, 1975, *AIChE J.* **21**: 510.

Leland, T. W. and P. S. Chappelear, 1968, *Ind. Eng. Chem.* **60(7)**: 15.

Lemmon, E. W. and R. T. Jacobsen, 2000, *J. Phys. Chem. Ref. Data* **29**: 521.

Li, P. and Z. H. Xu, 1993, *J. Chem. Ind. Eng. China* **44**: 129.

Li, P., X. Y. Zheng, and J. F. Lin, 1991, *Fluid Phase Equil.* **67**: 173.

Liu, Y. F. and Z. Xu, 1991, *Fluid Phase Equil.* **66**: 263.

Luettmer-Strathmann, J., S. Tang, and J. V. Sengers, 1992, *J. Chem. Phys.* **97**: 2705.

Lyckman, E. W., C. A. Eckert, and J. M. Prausnitz, 1965, *Chem. Eng. Sci.* **20**: 703.

Machado, J. R. S. and W. B. Streett, 1983, *J. Chem. Eng. Data* **28**: 218.

Magee, J. W., 1996, *Int. J. Thermophys.* **17**: 803.

Magee, J. W., 1998, *Int. J. Thermophys.* **19**: 1381.

Malhotra, R. and L. A. Woolf, 1991, *J. Chem. Thermodyn.* **23**: 867.

Martin, J. J. and Y. C. Hou, 1955, *AIChE J.* **1**: 142.

Martin, J. J., R. M. Kapoor, and N. de Nevers, 1959, *AIChE J.* **5**: 159.

McClune, C. R., 1976, *Cryogenics* **16**: 289.

Michels, A., T. Wassenaar, G. J. Wolkers, C. H. R. Prins, and L. V. D. Klundert, 1966, *J. Chem. Eng. Data.* **11**: 449.

Mulla, K. and V. F. Yesavage, 1989, *Fluid Phase Equil.* **52**: 67.

Natour, G., H. Schuhmacher, and B. Schramm, 1989, *Fluid Phase Equil.* **49**: 67.

Nicoll, J. F. and J. K. Bhattacharjee, 1981, *Phys. Rev. B* **23**: 389.

Nicoll, J. F. and P. C. Albright, 1985, *Phys. Rev. B* **31**: 4576.

Nicoll, J. F., 1981, *Phys. Rev. A* **24**: 2203.

Nowak, P., R. Kleinrahm, and W. Wagner, 1996, *J. Chem. Thermodyn.* **28**: 1441.

Nowak, P., R. Kleinrahm, and W. Wagner, 1997, *J. Chem. Thermodyn.* **29**: 1157.

O'Connell, J. P. and J. M. Prausnitz, 1967, *Ind. Eng. Chem. Process Des. Dev.* **6**: 245.

Olf, G., A. Schnitzler, and J. Gaube, 1989, *Fluid Phase Equil.* **49**: 49.

Osborne, N. S., H. F. Stimson, and D. C. Ginnings, 1939, *J. Res. Natl. Bur. Stand.* **23**: 261.

Outcalt, S. L. and M. O. McLinden, 1996, *J. Phys. Chem. Ref. Data* **25**: 605.

Patel, N. C. and A. S. Teja, 1982, *Chem. Eng. Sci.* **37**: 463.

Van Pelt, A. and J. V. Sengers, 1995, *J. Supercrit. Fluids* **8**: 81.

Peneloux, A., E. Rauzy, and R. Freze, 1982, *Fluid Phase Equil.* **8**: 7.

Peng, D. Y. and D. B. Robinson, 1976, *Ind. Eng. Chem. Fund.* **15**: 59.

Piao, C. -C. and M. Noguchi, 1998, *J. Phys. Chem. Ref. Data* **27**: 775.

Pitzer, K. S. and R. F. Curl, 1957, *J. Am. Chem. Soc.* **79**: 2369.

Pitzer, K. S., D. Z. Lippmann, R. F. Curl, C. M. Huggins, and D. E. Petersen, 1955, *J. Am. Chem. Soc.* **77**: 3433.

Polak, J. and B. C. -Y. Lu, 1972, *Can. J. Chem. Eng.* **50**: 553.

Rackett, H. G., 1970, *J. Chem. Eng. Data* **15**: 514.

Redlich, O. and J. N. S. Kwong, 1949, *Chem. Rev.* **44**: 233.

Ree, F. H. and W. G. Hoover, 1964, *J. Chem. Phys.* **40**: 939.

Reid, R. C., J. M. Prausnitz, and B. E. Poling, 1987, The Properties of Gases and Liquids, 4th ed. (McGraw-Hill, New York).

Reid, R. C., J. M. Prausnitz, and T. K. Sherwood, 1977, The Properties of Gases and Liquids, 3rd ed. (McGraw-Hill, New York).

Riedel, L., 1954, *Chem. Ing. Tech.* **26**: 83.

Schloms, R. and V. Dohm, 1989, *Nuclear Phys. B* **328**: 639.

Schmidt, G. and H. Wenzel, 1980, *Chem. Eng. Sci.* **35**: 1503.

Setzmann, U. and W. Wagner, 1991, *J. Phys. Chem. Ref. Data.* **20**: 1061.

Smith, G. E., R. E. Sonntag, and G. J. Van Wylen, 1963, *Adv. Cryo. Eng.* **8**: 162.

Smith, G. E., R. E. Sonntag, and G. J. Van Wylen, 1964, *Adv. Cryo. Eng.* **9**: 45.

Smith, J. M., H. C. Van Ness, and M. M. Abbott, 1996, Introduction to Chemical Engineering Thermodynamics, 5th ed. (McGraw-Hill, New York).

Soave, G., 1972, *Chem. Eng. Sci.* **27**: 1197.

Span, R., E. W. Lemmon, R. T. Jacobsen, W. Wagner, and A. Yokozeki, 2000, *J. Phys. Chem. Ref. Data* **29**: 1361.

Span, R. and W. Wagner, 1996, *J. Phys. Chem. Ref. Data* **25**: 1509.

Spencer, C. F. and D. P. Danner, 1972, *J. Chem. Eng. Data* **17**: 236.

Spencer, C. F. and S. B. Adler, 1978, *J. Chem. Eng. Data* **23**: 82.

Srinivasan, K., 1989, *Int. J. Refrig.* **12**: 194.

Starling, K. E., 1972, *Hydrogcarbon Process* **51(5)**: 129.

Streatfeild, M. H., C. Henderson, L. A. K. Staveley, A. G. M. Ferreira, I. M. A. Fonseca, and L. Q. Lobo, 1987, *J. Chem. Thermodyn.* **19**: 1163.

Su, G. S., 1937, D. Sc. thesis, Massachussett Institute of Technology, Cambridge,

Massachussett, USA.

Sunaga, H., R. Tillner-Roth, H. Sato, and K. Watanabe, 1998, *Int. J. Thermophys.* **19**: 1623.

Tang, S., G. X. Jin, and J. V. Sengers, 1991, *Int. J. Thermophys.* **12**: 515.

Tegeler, Ch., R. Span, and W. Wagner, 1999, *J. Phys. Chem. Ref. Data* **28**: 779.

Teja, A. S., S. I. Sandler, and N. C. Patel, 1981, *Chem. Eng. J.* **21**: 21.

Ter-Gazarian, G., 1906, *J. Chim. Phys.* **4**: 140.

Thomas, R. H. P. and R. H. Harrison, 1982, *J. Chem. Eng. Data* **27**: 1.

Thomson, G. H., K. R. Brobst, and R. W. Hankinson, 1982, *AIChE J.* **28**: 671.

Tillner-Roth, R. and H. D. Baehr, 1992, *J. Chem. Thermodyn.* **24**: 413.

Tillner-Roth, R. and H. D. Baehr, 1994, *J. Phys. Chem. Ref. Data* **23**: 657.

Tillner-Roth, R. and A. Yokozeki, 1997, *J. Phys. Chem. Ref. Data* **26**: 1273.

Trebble, M. A. and P. R. Bishnoi, 1987, *Fluid Phase Equil.* **35**: 1.

Tsonopoulos, C, 1974, *AIChE J.* **20**: 263.

Tsonopoulos, C, 1975, *AIChE J.* **21**: 827.

Tsonopoulos, C. and D. Ambrose, 1995, *J. Chem. Eng. Data* **40**: 547.

Tsonopoulos, C. and J. H. Dymond, 1997, *Fluid Phase Equil.* **133**:11.

Tsonopoulos, C., J. H. Dymond, and A. M. Szafranski, 1989, *Pure Appl. Chem.* **61**: 1387.

Wagner, W., V. Marx, and A. Pruss, 1993, *Int. J. Refrig.* **16**: 373.

Wagner, W. and A. Pruss, 2002, *J. Phys. Chem. Ref. Data* **31**: 387.

Watson, K. M., 1943, *Ind. Eng. Chem.* **35**: 398.

Watson, P., M. Cascella, D. May, S. Salerno, and D. Tassios, 1986, *Fluid Phase Equil.* **27**: 35.

Weber, L. A., 1994, *Int. J. Thermophys.* **15**: 461.

Wilson, K. G., 1971, *Phys. Rev. B* **4**: 3174.

Wilson, K. G., 1983, *Rev. Mod. Phys.* **55**: 583.

Wu, G. Z. A. and L. I. Stiel, 1985, *AIChE J.* **31**: 1632.

Wyczalkowska, A. K., Kh. S. Abdulkadirova, M. A. Anisimov, and J. V. Sengers, 2000, *J. Chem. Phys.* **113**: 4985.

Wyczalkowska, A. K. and J. V. Sengers, 1999, *J. Chem. Phys.* **111**: 1551.

Xiang, H. W., 2001a, *Int. J. Thermophys.* **22**: 919.

Xiang, H. W., 2001b, *J. Phys. Chem. Ref. Data* **30**: 1161.

Xiang, H. W., 2002, *Chem. Eng. Sci.* **57**: 1439.

Xiang, H. W., 2003, Thermophysico-chemical Properties of Fluids: Corresponding -States Principle and Practise (Science Press, Beijing), in Chinese.

Yen, L. C. and S. S. Woods, 1966, *AIChE J.* **12**: 95.

Yergovich, T. W., G. W. Swift, and F. Kurata, 1971, *J. Chem. Eng. Data* **16**: 222.

Young, S., 1910, *Sci. Proc. Roy. Soc. Dublin*, **12**: 374.

Younglove, B. A. and M. O. McLinden, 1994, *J. Phys. Chem. Ref. Data.* **23**: 731.

Yu, J. M. and B. C. Y. Lu, 1987, *Fluid Phase Equil.* **34**: 1.

Zhang, B. J. and Y. C. Hou, 1987, *J. Chem. Ind. Eng. China* **4**: 445.

Zhang, B. J. and Y. C. Hou, 1989, *J. Chem. Ind. Eng. China* **6**: 263.

Chapter 6

Vapor Pressures

6.1 Introduction

The vapor pressure is required in developing equations of state, in studying first-order and second-order vapor-liquid phase transitions, in obtaining the other thermodynamic properties of substances, in deriving the enthalpy and entropy in the two-phase region, and in describing binary and multicomponent mixtures. As a result, a generalized vapor-pressure equation is of great significance in fundamental theories and in engineering applications. For these reasons, the objective of extensive studies based on both theoretical ideas and empirical approaches has been to describe the vapor pressure as a function of temperature along the entire vapor-liquid coexistence curve.

6.2 Phase Transition Theory

The Clausius-Clapeyron equation proposed in 1834 is one of the earliest fundamental contributions to physical chemistry. It is

$$\mathrm{d}p / \mathrm{d}T = \Delta H / T \Delta V \tag{6.1}$$

or

$$- \mathrm{d} \ln p / \mathrm{d}(1/T) = \Delta H / R \Delta Z , \tag{6.2}$$

which gives the exact thermodynamic relation between the saturated vapor pressure p, the latent heat or enthalpy of vaporization ΔH, the temperature T, and the volume change ΔV or the compressibility-factor change ΔZ accompanying vaporization. According to the modern understanding of critical phenomena from critical scaling laws, there is a theoretical value for not only theleading non-analytic critical exponent but also the corresponding amplitude; specifically, certain ratios of amplitudes are universal constants (Moldover, 1985, Moldover and Rainwater, 1988; Rainwater and Lynch, 1989).

Besides giving a reasonably precise representation of the vapor pressure, a vapor-pressure equation may entail suitable values for the enthalpy of vaporization of substances based on Eqs. (6.1) and (6.2). Waring (1954) indicated that the form of the curve for $\Delta H / R\Delta Z$ as a function of temperature provides a qualitative test for the suitability of a vapor-pressure equation. The region below the temperature of minimum $\Delta H / R\Delta Z$, T_{min}, which is for most substances at a reduced temperature of about between 0.80 and 0.85, the first derivative of $\Delta H / R\Delta Z$ must be negative and the second derivative positive. An equation that does not lead to this should in general not be used for extrapolation, although it may be used satisfactorily for interpolation among experimental data within their range. The third derivative should also be positive. At very low reduced temperatures the first derivative is still negative. The second and third are expected to be positive, but in practice may be positive, zero, or negative, as long as they are small compared to their values at higher temperatures. At temperatures above T_{min}, the first and second derivatives must both be positive and the third seems also to be positive, at least up to very near the critical temperature. Thus, at least three parameters are necessary when an equation is intended to represent these physical behaviors.

6.3 Vapor-Pressure Equation

6.3.1 Overview of Vapor-Pressure Equations

Most vapor-pressure estimations and correlation equations stem from an integration of Eqs. (6.1) or (6.2). When this is done, an assumption must be made regarding the dependence of the quantity $\Delta H / \Delta Z$ on temperature. The simplest approach is to assume that the quantity $\Delta H / \Delta Z$ is constant and independent of temperature. Then the Clapeyron equation is obtained:

$$\ln p = a_1 + a_2 / T . \tag{6.3}$$

This is a fairly good relation for approximating vapor pressures over small temperature intervals.

Antoine (1888) proposed a three-parameter simple modification of the Clapeyon vapor-pressure equation, which has been widely used in engineering practice, as follows:

$$\ln p = a_1 + a_2 /(T + a_3) . \tag{6.4}$$

The applicable range corresponds to an interval of about 1 to 200 kPa in pressure, and the equation cannot be applied to the high-pressure range. The Antoine equation is a useful approximate equation but does not represent vapor pressures to within experimental error. Cox (1936) proposed the following equation:

$$\ln p = (1 - a_1 / T)(a_2 + a_3 T + a_4 T^2)$$ (6.5)

where a_4 is another substance-dependent parameter. This equation is not accurate enough and does not have adequate extrapolation capacity.

Frost and Kalkwarf (1953) maintained the assumption of the linear relationship for ΔH but calculated the ΔZ value in a more rigorous way than the Van der Waals equation of state. The resulting equation is

$$\ln p = a_1 + a_2 / T + a_3 \ln T + a_4 p / T^2 .$$ (6.6)

However, this leads to a somewhat involved calculation and is responsible for the nonexplicit form.

Riedel (1954) applied two very rough approximations: namely, that ΔZ was equal to unity and that ΔH varies linearly with temperature. However, the inaccuracies introduced by these assumptions are reduced by a corrective term proportional to T_r^6

$$\ln p_r = a_1 + a_2 / T_r + a_3 \ln T_r + a_4 T_r^6 ,$$ (6.7)

where $T_r = T / T_c$ is the reduced temperature and T_c the critical temperature; $p_r = p / p_c$ is the reduced pressure and p_c the critical pressure.

Miller (1964) applied the empirical relations for ΔH by Watson (1948) and for ΔZ by Haggenmacher (1946) to describe the relation of vapor pressure and temperature:

$$\Delta H = \Delta H_b (1 - T_r)^n / (1 - T_{rb})^n$$ (6.8)

$$\Delta Z = (1 - P_r / T_r^3)^{1/2} ,$$ (6.9)

where ΔH_b and T_{rb} are the enthalpy of vaporization and reduced temperature at the normal boiling temperature. Both of these equations can be regarded as good approximations for these properties over a wide range of temperatures. However, after substitution of the two empirical relations into Eq.(6.2), the presence of the pressure in the ΔZ term prevents direct integration of Eq.(6.2), and recourse to numerical methods is unavoidable.

Thek and Stiel (1966) pointed out that the Riedel, Frost-Karkwarf, and Miller equations were inaccurate in the low-pressure range. They attempted to combine the effect of ΔH and ΔZ dependence on temperature by expanding the ratio $\Delta H / \Delta Z$ as a power series in reduced temperature truncated after the fourth term. Furthermore, a corrective term was added, which in turn is a function of reduced temperature only:

$$\ln p_r = a_1 f(T_r) + a_2 [(T_r^{a_3} - 1) / a_3 + 0.04(T_r^{-1} - 1)]$$

$$f(T_r) = 1.14893 - T_r^{-1} - 0.11719 T_r - 0.031714 T_r - 0.375 \ln T_r. \tag{6.10}$$

Their final equation is effective in calculating vapor pressures with satisfactory accuracy in both the low- and high-pressure regions. However, an obvious drawback is the dependence of the vapor pressure on ΔH at the normal boiling temperature, which appears as a proportionality factor in the series expansion.

Goodwin (1969) included nonanalytic behavior at the critical point and presented vapor-pressure equations for oxygen and nitrogen:

$$\ln(p / p_t) = a_1 x + a_2 x^2 + a_3 x^3 + a_4 x(1 - x)^{1.89}, \tag{6.11}$$

where $x = (1 - T_t / T)(1 - T_t / T_c)$. Eq. (6.11) could not be applied widely since only a few triple-point pressure of fluids are known, where p_t and T_t are the pressure and temperature at the triple point.

Ambrose et al. (1970) proposed that the Chebyshev polynomials may accurately describe the vapor pressure of fluids from the triple point to the critical point:

$$T \ln p = 0.5 a_0 + \sum a_s E_s(x) \tag{6.12}$$

$$E_s(x) = \cos(s \arccos x) \qquad x = [2T - (T_h - T_l)] / (T_h - T_l),$$

where a_0 and a_s are substance-dependent parameters, and T_h and T_l are the highest and lowest temperatures of the temperature interval, but this method contains several adjustable coefficients, even up to seven.

Wagner (1973) developed a four-parameter equation with a stepwise procedure that optimizes selection of a functional form of an equation for each substance:

$$\ln p_r = (a_1 \tau + a_2 \tau^{1.5} + a_3 \tau^3 + a_4 \tau^6) / T_r, \tag{6.13}$$

where $\tau = 1 - T_r$. The equation was initially derived to describe the vapor pressures of argon and nitrogen from the triple point to the critical temperature. This equation has been applied by Ambrose (1978), by Ambrose and Patel (1984), by Chase (1984), by McGarry (1983), by Reid et al. (1958, 1966, 1977, 1987), by Scott and Osborn (1979), and by Smith and Srivastava (1986). This equation can represent the vapor pressure over a wide range and generates a reasonable shape for the vapor pressure curve from a reduced temperature of 0.5 up to the critical point, but it diverges too strongly near the critical point. This equation may not extrapolate well to reduced temperatures below 0.5 (Reid et al., 1987).

Thomas (1976) proposed an equation based on the observation that the ratio of the value of $RT d \ln p / dT$ for any nonassociating compound to the value of the function for any other such compound at the same vapor pressure is constant over a range from a few millimeters of mercury to the critical pressure

$$p = 253312 / [\exp(x) - a_3]$$

(6.14)

$$x = a_1 + a_2 \ln T .$$

Xu (1984) used a polynomial of the third degree relating the latent heat to the reduced temperature, and ΔZ was assumed to be constant. The inaccuracies were reduced by a corrective term proportional to T_r^4:

$$\ln p = a_1 \ln T_r + a_2 (T_r - T_r^{-1}) + a_3 (T_r^2 - 4T_r^{-1} + 3) + a_4 (T_r^4 - 16T_r^{-1} + 15) .$$

(6.15)

Vetere (1986) proposed a four-parameter vapor-pressure equation,

$$\ln p_r = a_1 (1 - T_r^{-1}) - a_2 \ln T_r + a_2 (T_r^{-1} - 1) / a_3 (a_3 - 1)) T_f^{a_3 - 1} ,$$

(6.16)

where T_f is a substance-dependent parameter. The equation incorporates an inflection-point condition and has a simple relation, which describes the behavior with a minimal number of coefficients, and a polynomial of the third degree, by expressing the ratio $\Delta H / \Delta Z$ as a function of the reduced temperature only. However, ethanol and propanol are recognized exceptions.

Iglesias-Silva et al. (1987) proposed a vapor-pressure equation from the asymptotic scaling-law behavior and extension of the Churchill-Usagi correlation technique from heat transfer and fluid mechanics:

$$p = (p_c - p_t)[1 - p(y)] + p_t$$

(6.17)

where

$$p(y) = p_0 (y)^n + p_\infty (y)^n]^{1/n}$$

$$p_0(y) = a_0 + a_1(a_3y + 1)\exp(\frac{-a_2 + b_0/R}{a_3y + 1})$$

$$p_\infty(y) = 2 - a_4(1 - y) + a_5(1 - y)^{2-0.199} + a_6(1 - y)^3 + a_7(1 - y)^4.$$

The equation and the determination of its coefficients are quite complicated.
Xiang and Tan (1994) proposed a vapor-pressure equation as:

$$\ln p_r = (a_0 + a_1\tau^{n_1} + a_2\tau^{n_2})\ln T_r, \tag{6.18}$$

which (1) satisfies the renormalization-group theory of critical phenomena and possesses a simple and reasonable physical form.
(2) has high accuracy to produce experimental vapor-pressure data over the entire temperature range with only 3 parameters and 2 exponents. However, its capacity far exceeds that of the Antoine equation, which has the same number of adjustable parameters. Equation (6.18) confirms the theoretical analysis of Waring (1954) in that an equation to represent the physical behavior over the entire vapor-pressure range requires at least three adjustable parameters as shown in Figs. 6.1 and 6.2.
(3) has obvious physical behavior; two exponents are universal. One is $n_1 = 2 - \alpha = 1.89$, which is based on the renormalization-group theory; the other is $n_2 = 3n_1 = 5.67$, which is 3 times n_1. One of the parameters is the Riedel factor and the other two have good regularity.
(4) overcomes the other empirical shortcomings of the previous vapor-pressure equations and has an excellent prediction capacity from the usual range in which data are available both to the critical point and critical parameters, and in particular, to the low temperature range below the reduced temperature of 0.5, in contrast to the other existing equations.
(5) has good generalization, and is applicable to simple, quantum, nonpolar, polar, hydrogen-bonding, and associating molecules.

6.3.2 Calculation of Vapor Pressure

The fitted accuracy to a data set is dependent on the adjustable parameters of the corresponding vapor-pressure equation. A three-parameter equation is sufficient and suitable to use when simple calculation is needed. The Antoine equation was common for this purpose earlier, and the Xiang-Tan equation is available now.
The Xiang-Tan equation was applied to a diverse set of substances including simple, nonpolar, polar, quantum, hydrogen-bonding, and associating substances,

which have been experimentally investigated in detail to obtain reliable experimental data (Xiang and Tan, 1994). The sources of these experimental data are listed in Table 6.1. The three parameters of some substances are revised and presented in Table 6.2, along with the critical temperature and critical pressure. A comparison has been made between the vapor-pressure values calculated from the Xiang-Tan equation and the experimental data (Xiang and Tan, 1994). Considering the uncertainties of the experimental data, the Xiang-Tan equation is accurate in the entire range from the triple point to the critical point. The Xiang-Tan equation provides accuracy comparable to that of the Wagner equation in representing experimental data.

For example, in Table 6.3 and Fig. 6.1, the calculated results from the Antoine equation and the Xiang-Tan equation are presented for water, for which highly accurate experimental data are available. Tolerances of the critical temperature and the critical pressure given by IAPWS (1994) are 0.015% and 0.15%, respectively.

As a vapor-pressure equation, Eq. (6.18) is consistent with the renormalization-group theory with regard to the exponent, but not the amplitude. On expanding Eq. (6.18) in a power series in τ, the amplitude of the leading-order nonanalytic term is not consistent with the correct universal constant (Moldover and Rainwater, 1988; Rainwater and Lynch, 1989), which has a very weak influence on the second derivative and becomes significant only for less than several thousandths of τ, when the temperature approaches the critical point, as can be seen in Table 6.4 for water. The Xiang-Tan equation, which has only three substance-dependent parameters, can reproduce accurate experimental data from the triple point to the critical point, as can be seen in Table 6.3 and Fig. 6.1, and basically conforms to the behavior of $\Delta H / R\Delta Z$ and its derivatives as a function of temperature as can be seen in Fig. 6.2 and Table 6.4. The Wagner equation diverges too strongly in terms of $\tau^{-0.5}$ (actually it should be only $\tau^{-0.11}$). As can be seen in Table 6.4, the second derivative of pressure calculated from the Wagner equation is too large, while the second derivative from the Xiang-Tan equation is slightly smaller than that from the critical power law near the critical point.

Table 6.1 Source of Experimental Data

Substance	Refs.
Argon	Gilgen et al., 1994
Nitrogen	Nowak et al., 1997
Methane	Kleinrahm and Wagner, 1986
Carbon dioxide	Duschek et al., 1990
Ethylene	Nowak et al., 1996
Ethane	Thomas, 1976; Stray and Tsumura, 1976, Pal et al., 1976
Propane	Thomas, 1976; Kratzke and Mueller, 1984; Thomas

	and Harrison, 1982; Kemp and Egan, 1938; Carruth and Kobayashi, 1973
Benzene	Ambrose et al., 1981; Connolly and Kandalic, 1962
Decane	Ruzicka and Majer, 1994; Chirico et al., 1989
Dichlorodifluoro-methane	Blanke and Weiss, 1992a; Fernandez-Fassnacht and del Rio, 1985; Gilkey et al., 1931; Gordon and McWilliam, 1946; Haendel et al., 1992; Michels et al., 1966; Mollerup and Fredenslund, 1976
Chlorodifluoromethane	Benning and McHarness, 1940; Blank and Weiss, 1992a; Booth and Swinehart, 1935; Giuliani et al., 1995; Goodwin et al., 1992; Haendel et al., 1992; Hongo et al., 1990; Neilson and White, 1957
Carbon terachloride	Boublik et al., 1984
Chloroform	Boublik et al., 1984; Campbell and Chatterjee, 1968
Trifluoromethane	Hori et al., 1981; Hou and Martin, 1959; Popowicz et al., 1982; Stein and Proust, 1971; Valentine et al., 1962; Varshni and Mohan, 1954
Dichloromethane	Boublik et al., 1984
Methyl chloride	Beersmans and Jungers, 1947; Holldorff and Knapp, 1988; Hsu and McKetta, 1964; Mansoorian et al., 1981; Messerly and Aston, 1940
Methyl fluoride	Biswas et al., 1989; Bominaar et al., 1987; Collie, 1889; Demiriz et al., 1993; Fonseca and Lobo, 1994; Grosse et al., 1940; Michels and Wassenaar, 1948; Oi et al., 1983
Difluoromethane	De Vries, 1997; Magee, 1996; Fu et al., 1995; Defibaugh et al., 1994; Weber and Silva, 1994; Weber and Goodwin, 1993; Malbrunot et al., 1968
2,2-Dichloro-1,1,1-Trifluoroethane	Goodwin et al., 1992a
Pentafluoroethane	De Vries, 1997; Magee, 1996; Weber and Silva, 1994; Tsvetkov et al., 1995; Boyes and Weber, 1994
1,1,1,2-tetrafluoroethane	Magee and Howley, 1992; Goodwin et al., 1992; Wilson and Basu, 1988; Blanke et al., 1995; Weber, 1989; Baehr and Tillner-Roth, 1991
1,1-difluoroethane	Wilson and Basu, 1988; Blanke and Weiss, 1992a; Silva and Weber, 1993
Bromobenzene	Boublik et al., 1984; Stull, 1947
Chlorobenzene	Boublik et al., 1984
Hydrogen cyanide	Boublik et al., 1984
Formaldehyde	Boublik et al., 1984; Spence and Wild, 1935
Methyl mercaptan	Boublik et al., 1984
Methylamine	Stull, 1947; Aston et al., 1937

Acetonitrile	Mousa, 1981; Kratzke and Mueller, 1985
1,1-Dichloroethane	Li and Pitzer, 1956
1,2-Dichloroethane	Boublik et al., 1984
Acetaldehyde	Boublik et al., 1984
Dimethyl ether	Boublik et al., 1984; Kennedy et al., 1941
Dimethyl sulfide	Boublik et al., 1984; Osborne et al., 1942
Ethyl mercaptan	Boublik et al., 1984
Ethyl formate	Stull, 1947
Methyl acetate	Boublik et al., 1984; Stull, 1947
Acetone	Ambrose et al., 1974
Ammonia	Baehr et al., 1976; Zander and Thomas, 1979; Cragoe et al., 1920
Water	IAPWS, 1994
Methanol	Ambrose et al., 1975; Cooney and Morcom, 1988; Gibbard and Greek, 1974; Kretschmer and Wiebe, 1949
Ethanol	Ambrose et al., 1970; Lydersen and Tsochev, 1990; Ambrose et al., 1975
1-Propanol	Ambrose et al., 1975; Lydersen and Tsochev, 1990; Ambrose and Townsend, 1963
Isopropyl alcohol	Ambrose et al., 1975; Kretschmer and Wiebe, 1952
n-Propylamine	Stull, 1947; Boublik et al., 1984
Ethyl acetate	Ambrose et al., 1967
Methyl ethyl ketone	Ambrose, 1981
n-Butyl alcohol	Ambrose et al., 1975; Ambrose and Townsend, 1963
Acetic acid	Ambrose et al., 1977; Potter and Ritter, 1954
Isobutyric acid	Stull, 1947
Methyl proptonate	Boublik et al., 1984
Ethyl ether	Ambrose et al., 1974; Taylor and Smith, 1922
Diethylamine	Stull, 1947
Methyl isopropyl ketone	Ambrose et al., 1981
2-Pentanone	Ambrose et al., 1981
Phenol	Boublik et al., 1984
Aniline	Boublik et al., 1984
Methyl isobutyl ketone	Boublik et al., 1984
Diisopropyl ether	Boublik et al., 1984
m-Cresol	Boublik et al., 1984
o-Cresol	Boublik et al., 1984
p-Cresol	Boublik et al., 1984

Table 6.2 Critical Parameters and Xiang-Tan Vapor-Pressure Parameters for Some Substances [Adapted from Xiang and Tan (1994)]

Substance	T_c (K)	p_c (kPa)	a_0	a_1	a_2
Argon	150.69	4865	5.78471374	6.24383974	12.3718357
Oxygen	154.581	5043	5.91231393	6.54549694	12.5794134
Nitrogen	126.2	3400	5.9844923	6.76437425	14.681488
Chlorine	416.9	7972	6.2303934	7.47680282	16.8668022
Carbon dioxide	304.136	7377	6.84599542	10.2023639	9.35969257
Methane	190.551	4598	5.87304544	6.23280143	13.0721578
Ethane	305.33	4872	6.30717658	7.47042131	17.0958137
Propane	369.80	4239	6.50580501	8.6776247	18.0116214
Butane	425.2	3800	6.81692028	8.77671813	23.7680492
2-Methylbutane	460.95	3390	6.92686796	9.26964282	22.4417114
Cyclohexane	553.64	4075	6.81255435	9.41438293	23.2468109
Perfluorotoluene	534.5	2710	8.1846466	12.7024135	32.9377937
Benzene	561.75	4875	6.82740545	9.34241485	24.1741504
Toluene	591.72	4105	7.12600946	9.81328392	24.3763809
Ethylbenzene	617.12	3605	7.31366586	10.4335994	26.1998558
o-Xylene	630.25	3733	7.37496614	10.3818836	25.9129962
m-Xylene	616.97	3536	7.44022798	10.6410732	25.2128753
p-Xylene	616.15	3511	7.44173097	10.3619289	26.6721134
Neon	44.448	2664	5.64703512	5.33091354	12.2967739
Hydrogen	33.19	1315	4.78105258	2.82220339	2.87949132
Helium	5.2014	227.5	3.83707499	1.16239106	2.63104414
Trichlorofluoromethane	471.15	4487	6.83648252	8.45433425	22.0910034
Dichlorodifluoromethane	385	4140	6.70782089	8.77115631	18.4986419
Chlorodifluoromethane	369.3	4988	6.88100147	9.47825336	20.3845615
Difluoromethane	351.56	5828	7.25783849	9.44906806	21.9283714
Trichlorotrifluoroethane	487.483	3410	7.03479385	9.96937179	22.8604393
2,2-Dichloro-1,1,1-Trifluoroethane	456.87	3665	7.17751407	10.4138793	23.9151554
1,1,2,2-Tetrafluoroethane	392	4640	7.253801	10.62278	24.97116
1-Chloro-1,1-difluoroethane	410.29	4041	6.93208	9.567646	22.21399
1,1,1,2-tetrafluoroethane	374.18	4056	7.41535425	10.9633932	23.8982505
1,1-difluoroethane	386.44	4500	7.184458	9.936063	22.058460

Ethyl fluoride	375.31	5027	6.9000082	9.12291336	16.228197
Diethy ether	466.74	3637	7.1713705	10.3994998	24.3123378
Water	647.14	22050	7.60794067	10.1932439	21.1083545
Ammonia	405.5	11350	7.11388492	9.51535415	19.5376377
Nitromethane	588	6310	7.96073532	6.91034173	51.2841186
TFE	498.53	4800	8.63110256	19.050188	14.7720022
Acetic acid	592.71	5786	8.38445949	9.78763198	29.3230476
Pentafluorochloroa cetone	410.65	2877	7.45677614	11.620016	26.5720901
Acetone	508.1	4696	7.37726593	10.0870218	23.4207172
Hexafluoroacetone	357.25	2841	7.51933526	12.1870098	24.4911994
Methanol	511	8130	8.937361	12.29697	24.32896
Ethanol	513.92	6148	9.224126	14.96870	35.57417
1-Pentanol	588.2	3910	8.18587684	18.9232025	53.7686767

Table 6.3 Vapor Pressure of Water from the Xiang-Tan Equation (6.18) and Antoine Equation (6.4) (IAPWS, 1994)

T_c (K)	T_r $= T / T_c$	p (MPa)	$p_r = p / p_c$	$100(1 - p_{calc} / p_{expt})$ Xiang-Tan Eq. (6.18)	$100(1 - p_{calc} / p_{expt})$ Antoine Eq. (6.4)
273.160	0.422	6.12E-04	2.77E-05	0.013	0.120
278.149	0.430	8.72E-04	3.96E-05	0.016	0.016
283.147	0.438	1.23E-03	5.57E-05	0.019	-0.039
288.146	0.445	1.71E-03	7.74E-05	0.011	-0.069
293.145	0.453	2.34E-03	1.06E-04	0.001	-0.074
298.143	0.461	3.17E-03	1.44E-04	-0.002	-0.057
303.142	0.468	4.25E-03	1.93E-04	-0.013	-0.038
308.141	0.476	5.63E-03	2.55E-04	-0.021	-0.015
313.140	0.484	7.38E-03	3.35E-04	-0.027	0.010
318.138	0.492	9.59E-03	4.35E-04	-0.025	0.036
323.137	0.499	1.23E-02	5.60E-04	-0.027	0.050
328.136	0.507	1.58E-02	7.15E-04	-0.027	0.056
333.134	0.515	1.99E-02	9.04E-04	-0.020	0.056
338.133	0.523	2.50E-02	1.14E-03	-0.016	0.004
343.132	0.530	3.12E-02	1.41E-03	-0.011	0.012
348.131	0.538	3.86E-02	1.75E-03	-0.005	-0.030
353.129	0.546	4.74E-02	2.15E-03	0.005	-0.081
358.128	0.553	5.78E-02	2.62E-03	0.011	-0.15
363.127	0.561	7.01E-02	3.18E-03	0.018	-0.23
368.126	0.569	8.45E-02	3.83E-03	0.023	-0.32
373.124	0.577	0.10132	4.60E-03	0.032	-0.43
383.123	0.592	0.14324	6.50E-03	0.033	-0.68

393.122	0.608	0.19848	9.00E-03	0.034	-0.98
398.120	0.615	0.23201	1.05E-02	0.037	-1.1
403.119	0.623	0.27002	1.22E-02	0.038	-1.3
413.117	0.638	0.36119	1.64E-02	0.031	-1.7
423.115	0.654	0.47571	2.16E-02	0.021	-2.1
433.113	0.669	0.61766	2.80E-02	0.010	-2.5
443.112	0.685	0.79147	3.59E-02	-0.005	-2.9
448.112	0.692	0.8918	4.05E-02	-0.014	-3.1
453.111	0.700	1.0019	4.54E-02	-0.018	-3.3
463.110	0.716	1.2541	5.69E-02	-0.030	-3.7
473.110	0.731	1.5537	7.05E-02	-0.037	-4.1
483.110	0.747	1.9062	8.65E-02	-0.044	-4.5
493.110	0.762	2.3178	0.105	-0.047	-4.9
503.110	0.777	2.7950	0.127	-0.042	-5.3
513.110	0.793	3.3446	0.152	-0.030	-5.6
523.110	0.808	3.9735	0.180	-0.017	-5.9
533.110	0.824	4.6892	0.213	-0.001	-6.1
543.111	0.839	5.4996	0.249	0.016	-6.3
553.111	0.855	6.4127	0.291	0.030	-6.5
563.111	0.870	7.4375	0.337	0.043	-6.6
573.111	0.886	8.5813	0.389	0.027	-6.7
583.111	0.901	9.8597	0.447	0.044	-6.6
593.111	0.917	11.278	0.512	0.032	-6.5
603.110	0.932	12.851	0.583	0.011	-6.3
613.110	0.947	14.593	0.662	-0.021	-6.0
623.109	0.963	16.521	0.749	-0.045	-5.5
633.108	0.978	18.657	0.846	-0.050	-4.9
643.107	0.994	21.033	0.954	0.007	-4.1
644.107	0.995	21.286	0.966	0.021	-4.0
645.107	0.997	21.542	0.977	0.037	-3.9
646.106	0.998	21.802	0.989	0.06	-3.8
647.10	1	22.045	1	0	

T_c=647.1 K p_c=22.045($\pm$0.003) MPa (Levelt Sengers et al., 1983)

a_0=7.603029 a_1=10.21238 a_2=21.0805

Average deviation=0.024%, Root-mean-square deviation=0.029%, Maximum deviation=0.060%

Table 6.4 First and Second Derivatives of Pressure for Water from the Xiang-Tan Equation, Critical Power Law, and Wagner Equation in the Vicinity of Critical Point

T(K)	p (MPa)	dp/dT (MPa/K)	$d^2 p/dT^2$ (MPa/K^2)		
			Xiang-Tan Eq. (6.18) in $\tau^{1.89}$ (0-order)	Critical Power Law in $\tau^{1.89}$ (-0.11-order)	Wagner Eq. (6.13) in $\tau^{1.5}$ (-0.5-order)
603.15	12.8568	0.1659	1.70E-03		
608.15	13.7078	0.1746	1.78E-03		
613.15	14.6034	0.1837	1.87E-03		
618.15	15.5456	0.1933	1.96E-03		
623.15	16.5368	0.2033	2.05E-03	2.03E-03	2.04E-03
628.15	17.5793	0.2138	2.16E-03		
633.15	18.6758	0.2249	2.27E-03	2.25E-03	2.31E-03
638.15	19.8291	0.2365	2.39E-03		
639.15	20.0668	0.2389	2.42E-03		
640.15	20.3070	0.2414	2.44E-03		
641.15	20.5496	0.2438	2.47E-03		
642.15	20.7946	0.2463	2.50E-03		
643.15	21.0422	0.2488	2.53E-03	2.57E-03	2.88E-03
644.15	21.2923	0.2514	2.55E-03	2.61E-03	3.03E-03
645.15	21.5449	0.2539	2.58E-03	2.67E-03	3.25E-03
646.15	21.8001	0.2565	2.61E-03	2.75E-03	3.72E-03

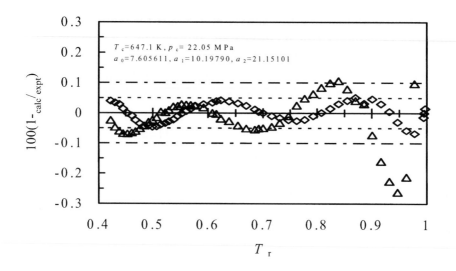

Fig. 6.1 Comparison of the vapor pressure and the derivative of pressure for water recommended from IAPWS (1994) with values calculated from the vapor pressure equation, Eq. (6.18). (◊), p ; (△), dp/dT.

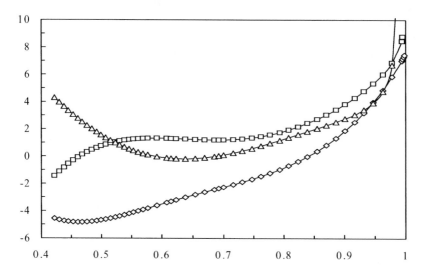

Fig 6.2 The derivatives of $\Delta H / R \Delta Z$ as a function of $T_r = T/T_c$ for water calculated from the Xiang-Tan Equation (6.18) (◊) the first derivative; (□) the second derivative; (△) the third derivative

6.3.3 Prediction and Extrapolation from Vapor-Pressure Equation

Besides a simple form and high accuracy, good extrapolation and accurate prediction are expected for a vapor-pressure equation. The vapor-pressure values at low pressure region may be obtained by extrapolation because the experimental data at lower pressures are rarely measured and usually are not accurate. If the substance has a high boiling temperature, its high-temperature experimental data are usually not known or they may be experimentally inaccessible because of decomposition. In consideration of these facts, extrapolation and prediction become important features of vapor-pressure equations, and they represent the cardinal principles by which the superiority of the corresponding equation is evaluated.

Reid et al. (1987) reviewed the vapor-pressure equations from the capacity of correlation and prediction over a wide range of vapor pressure. The three-parameter Antoine equation cannot correctly represent vapor-pressure behavior over an entire temperature range and, as a result, cannot be used for accurate prediction. The Wagner equation may reasonably represent the vapor-pressure from a reduced temperature of 0.5 up to the critical point; however, the equation may not be extrapolated well to reduced temperatures below 0.5 (Reid et al., 1987). The Xiang-Tan equation may accurately represent the vapor-pressure behavior over the entire range and also yields an excellent extrapolation from the usual range in which data are available to the triple point. A comparison of the extrapolation of the Wagner equation and Xiang-Tan equation was done by Xiang and Tan (1994). Since small absolute experimental error may result in a large relative error in the low pressure range, the extrapolated results in the low-temperature range displayed a relatively great deviation.

An example is shown in Fig. 6.3, in which the Xiang-Tan equation is extrapolated to predict the critical pressure of water from the experimental data between room temperature and the normal boiling point. The result is within 1% for the predicted critical pressure for water. It should be noted that the Wagner equation would generate over 30% deviation, which is not shown in Fig. 6.3.

The Xiang-Tan equation, with a simple physical form over the entire temperature range from the triple point to the critical point and only three adjustable parameters and two exponents, provides accuracy comparable to that of the Wagner equation, which has four adjustable parameters and four exponents. Since the formulation is based upon the physical behavior, the exponents and adjustable parameters reflect the physical properties of the substances. Incorporating and combining these physical associations, the Xiang-Tan equation can provide good predictive as well as correlative capabilities. The Xiang-Tan equation provides an accurate correlation and excellent extrapolation of the vapor-pressure behavior of simple, nonpolar, polar, quantum, hydrogen-bonding, and associating compounds.

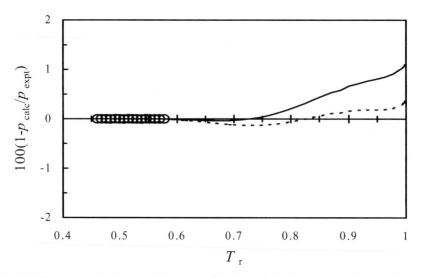

Fig. 6.3 The extrapolation results to predict the critical pressure of water based on the vapor-pressure equation, Eq. (6.18), whose substance-dependent coefficients a_0, a_1, a_2, and the critical pressure p_c are fitted only from the data in the temperature range between room temperature and the normal boiling point 373.124 K. (○) These data points used to fit Eq. (6.18): (—) The data between 298.15 K and the normal boiling point 373.124 K ($T_r = 0.577$) are used to fit a_0, a_1, a_2, and the critical pressure p_c in Eq. (6.18), these fitted parameters are then used to calculate the extrapolation deviations beyond the normal boiling point $T_r = 0.577$ up to the critical temperature of 647.1 K ($T_r = 1$); (---) similarly, the data between 293.15 K and 373.124 K are used.

6.4 CORRESPONDING-STATES VAPOR PRESSURE

6.4.1 Vapor Pressure from the Extended Corresponding-States Theory

To represent experimental data over the entire range from the triple point to the critical temperature, the vapor-pressure curve was based upon an equation with known physical behavior and had three substance-dependent parameters as described by Xiang and Tan and presented by Xiang (2001a,b, 2002).

The functional form presented in Eq. (6.18) is rewritten as

$$p_r = \exp\left[(a_0 + a_1 \tau^{n_1} + a_2 \tau^{n_2}) \ln T_r\right],$$ (6.19)

and

$$p = p_c \exp\left[(a_0 + a_1\tau^{n_1} + a_2\tau^{n_2})\ln T_r\right].$$ (6.20)

The slope of the vapor-pressure equation is given by

$$dp/dT = \frac{d\ln p_r}{dT_r}\, p/T_c,$$ (6.21)

in which

$$\frac{d\ln p_r}{dT_r} = (a_0 + a_1\tau^{n_1} + a_2\tau^{n_2})/T_r - (n_1 a_1\tau^{n_1-1} + n_2 a_2\tau^{n_2-1})\ln T_r.$$ (6.22)

At the critical point, the Riedel parameter α_c is:

$$\alpha_c = d\ln p/d\ln T.$$ (6.23)

The second derivative of pressure as a function of temperature is given by

$$d^2 p/dT^2 = \left[\frac{d^2\ln p_r}{dT_r^2} + (\frac{d\ln p_r}{dT_r})^2\right] p/T_c^2,$$ (6.24)

in which

$$\frac{d^2\ln p_r}{dT_r^2} = \left[n_1(n_1-1)a_1\tau^{n_1-2} + n_2(n_2-1)a_2\tau^{n_1-2}\right]\ln T_r$$

$$- (a_0 + a_1\tau^{n_1} + a_2\tau^{n_2})/T_r^2 - 2(n_1 a_1\tau^{n_1-1} + n_2 a_2\tau^{n_2-1})/T_r$$ (6.25)

Equation (6.18) can be used to correlate and extrapolate precise vapor pressure measurements at modest pressures to the critical point and to the triple point:

$$\ln p_r^{(0)} = (a_{00} + a_{10}\tau^{n_1} + a_{20}\tau^{n_2})\ln T_r$$

$$\ln p_r^{(1)} = (a_{01} + a_{11}\tau^{n_1} + a_{21}\tau^{n_2})\ln T_r$$

$$\ln p_r^{(2)} = (a_{02} + a_{12}\tau^{n_1} + a_{22}\tau^{n_2})\ln T_r.$$ (6.26)

The general coefficients a_{ij} of Eq. (6.26), given in Table 6.5, were found from fitting the vapor-pressure data for argon, the weakly nonspherical molecules of ethane, propane, difluoromethane, 1,1,1,2-tetrafluoroethane, and 1,1-difluoroethane, and the highly nonspherical water molecule. The coefficients are independent of the specific substance and are expected to be universal for all classes of molecules (Xiang, 2001a,b, 2002).

The updated critical parameters and corresponding-states parameters for some substances are listed in Table 6.6.

Table 6.5 General Coefficients of Eq. (6.26) (Xiang, 2001a,b, 2002)

a_{00}	5.790206	a_{10}	6.251894	a_{20}	11.65859
a_{01}	4.8888195	a_{11}	15.08591	a_{21}	46.78273
a_{02}	33.91196	a_{12}	- 315.0248	a_{22}	- 1672.179

Table 6.6 Molar Mass M, Critical Temperature T_c, Critical Pressure p_c, Critical Density ρ_c, Acentric Factor ω, and Aspherical Factor θ

Substance	M ($\mathrm{kg.kmol^{-1}}$)	T_c (K)	p_c (kPa)	ρ_c ($\mathrm{kg.m^{-3}}$)	ω	θ
Argon	39.948	150.69	4863	535	0.000	0.000
Nitrogen	28.013	126.19	3395	313	0.037	0.000
Methane	16.043	190.56	4598	162	0.011	0.007
Carbon dioxide	44.010	304.13	7377	468	0.225	0.245
Ethylene	28.054	282.35	5041	215	0.086	0.096
Ethane	30.070	305.32	4872	206	0.099	0.097
Propane	44.097	369.83	4248	220	0.152	0.171
Benzene	78.114	562.05	4895	305	0.210	0.472
Decane	142.86	617.70	2110	228	0.489	1.061
Trifluorobromo-methane	148.91	340.2	3970	750	0.173	0.128
Trichlorofluoro-methane	137.368	471.2	4408	560	0.187	0.196
Dichlorodifluoro-methane	120.91	384.95	4100	560	0.178	0.180
Chlorotrifluoro-methane	104.459	301.9	3890	610	0.178	0.606
Chlorodifluoro-methane	86.470	369.28	4988	525	0.221	0.501
Trifluoromethane	70.014	298.9	4800	530	0.263	1.21
Difluoromethane	52.020	351.26	5780	430	0.277	2.560
Chloromethane	50.488	416.25	6710	365	0.153	0.476
Fluoromethane	34.033	317.42	5880	310	0.197	2.06
Tetrachloromethane	153.823	556.3	4560	558	0.193	0.332

Chloromethane	119.378	536.0	5480	495	0.218	0.043
Dichloromethane	84.933	510.2	6080	448	0.199	0.334
2,2-Dichloro-1,1,1-trifluoroethane	152.93	456.83	3662	550	0.282	0.481
Pentafluoroethane	120.02	339.17	3620	570	0.306	0.388
1,1,1,2-Tetrafluoro-ethane	102.03	374.18	4055	511	0.327	0.885
1,1-Difluoroethane	66.050	386.41	4516	369	0.275	1.474
1,1-Dichloroethane	98.959	523.0	5070	440	0.240	0.771
1,2-Dichloroethane	98.959	561.0	5370	450	0.278	1.356
Ammonia	17.031	405.37	11345	234	0.256	2.028
Water	18.015	647.10	22050	325	0.344	3.947
Methanol	32.042	512.0	8000	270	0.560	4.486
Ethanol	46.069	513.9	6140	276	0.644	2.514
1-Propanol	60.096	536.5	5170	274	0.630	1.292
Hydrogen cyanide	27.026	456.6	5390	205	0.405	10.57
Formaldehyde	30.026	408.0	6590	260	0.280	4.311
Methyl mercaptan	48.109	469.7	7190	330	0.153	0.466
Methylamine	31.057	430.0	7430	210	0.292	0.301
Acetonitrile	41.053	545.4	4835	225	0.338	9.151
Acetaldehyde	44.054	461.0	5570	285	0.303	4.274
Acetic acid	60.052	593.0	5786	330	0.450	5.844
Dimethyl ether	46.069	400.0	5330	270	0.197	0.274
Dimethyl sulfide	62.136	503.0	5530	309	0.191	0.581
Ethyl mercaptan	62.136	499.0	5490	300	0.190	0.254
Acetone	58.080	508.1	4700	270	0.308	2.569
Ethyl formate	74.079	508.5	4740	320	0.285	0.928
Methyl acetate	74.079	506.8	4690	325	0.326	1.318
Isopropyl alcohol	60.096	508.0	4780	271	0.665	1.524
n-Propylamine	59.111	496.0	4810	254	0.300	0.345
Ethyl acetate	88.106	523.2	3880	300	0.367	0.787
Methyl ethyl ketone	72.107	536.8	4210	270	0.322	1.451
n-Butyl alcohol	74.123	563.0	4420	268	0.593	0.832
Isobutyric acid	88.107	609.0	4050	315	0.620	4.393
Methyl proptonate	88.107	530.6	4000	310	0.350	1.044
Ethyl ether	74.123	466.7	3640	258	0.281	0.420
Diethylamine	73.138	496.6	3708	242	0.305	0.346
Methyl isopropyl ketone	86.134	553.3	3850	282	0.332	1.182
2-Pentanone	86.134	561.5	3690	285	0.346	2.614
Bromobenzene	157.01	670.0	4520	484	0.251	0.717
Chlorobenzene	112.56	632.3	4520	365	0.250	0.618
Phenol	94.113	694.3	6100	400	0.438	1.712

Aniline	93.128	699.0	5350	340	0.384	1.433
Methyl isobutyl ketone	100.16	571.0	3300	280	0.385	1.710
Diisopropyl ether	102.18	500.0	2850	265	0.335	0.659
m-Cresol	108.14	705.8	4560	342	0.454	1.962
o-Cresol	108.14	697.4	5005	375	0.435	1.688
p-Cresol	108.14	704.6	5150	388	0.505	2.024

6.4.2 Comparison of Experimental Data and Existing Methods

For most cases as shown in Figs. 6.4 to 6.6, the extended corresponding-states theory for the representation of recent accurate vapor-pressure data for various typical real fluids over their entire temperature ranges shows that the agreement between the present method and the data is generally comparable to the agreement between the different data sets for each substance.

For highly accurate vapor-pressure data above the normal boiling point, the deviations of the calculated values from the experimental data are generally within 0.1%, which means the method is very useful to predict vapor pressure for those substances with high boiling temperatures, or where high-temperature experimental data are not known, or may be experimentally inaccessible because of decomposition. In the low-temperature region, the vapor pressure is subject to large uncertainties, especially at very low temperatures. Usually, the experimental uncertainties for the vapor pressure may be 5% to 10% at low temperatures. The extended corresponding-states theory accurately predicts the vapor pressure even at reduced temperatures as low as 0.3 to 0.2 for ethane and propane as shown in Fig. 6.4.

In contrast to existing models, the extended corresponding-states theory is able to describe the vapor pressure of highly polar, hydrogen-bonding and associating substances, such as dichlorodifluoromethane, chlorodifluoromethane, difluoromethane, 1,1-dichloro-2,2,2-trifluoroethane, pentafluoroethane, 1,1,1,2-tetrafluoroethane, 1,1-difluoroethane, water, methanol, ammonia, acetone, and acetonitrile as shown in Figs. 6.4 and 6.5 (Xiang, 2001a,b, 2002).

Typical substances such as trichlorofluoromethane, dichlorodifluoromethane, chlorotrifluoromethane, difluorochloromethane, trifluoromethane, and fluoromethane are shown in Figs. 6.6 to 6.11.

As shown by Xiang (2001a,b, 2002), the methods proposed by Lee and Kesler (1975), by Ambrose and Patel (1984), by Gupte and Daubert (1985), and by Edalat et al. (1993) do not have the prediction accuracy of this corresponding-states method which agrees well with highly accurate vapor pressure data in the high temperature range and with less accurate data at low temperatures. The results show that the extended corresponding-states method is substantially superior to conventional linear or quadratic corresponding-states correlations. It also proves that a quadratic correlation for the acentric factor is not a good representation of the vapor pressure for all classes of fluids. For polar substances, the other four-parameter corresponding-states models by Wilding and Rowley (1986) and by Wu and Stiel (1985) require critical parameters such as the radius of gyration or the acentric factor, and a pressure-density-temperature data point, which depends on a specific property. The results of the extended corresponding-states theory are also better than those obtained from any currently available method for a wider range of polar molecules.

Wilding et al. (1986) indicated that prediction results for alcohols are not very good for all models, and there are questions about the critical parameters and the reliability of experimental data because of decomposition. It can be seen that the extended corresponding-states method provides good prediction accuracy and is superior to commonly used methods.

As discussed by Xiang (2001a), by means of the extended corresponding-states theory, one may also determine critical parameters from available experimental data, which are usually limited approximately to the range 10 to 150 kPa. That is, if the property of a substance is some function of the reduced temperature, reduced pressure, and/or reduced density, then by knowing the values of the property over a range of temperatures, pressures, and/or densities, it is possible, in principle, to estimate the critical parameters and the acentric factor from this general corresponding-states principle. In principle, only several (not less than three) precise vapor-pressure data points spanning a wide range in the vapor-pressure curve are required to establish the three equations, using $f(T_c, p_c, \rho_c) = 0$, which have to be solved to obtain the three critical parameters T_c, p_c, and ρ_c. All of the other coefficients or parameters are known in the general corresponding-states theory, Eqs. (6.18) and (6.26), if we assume that the extended corresponding-states theory is applicable to the substance, at least to a reasonable degree. For example, a change of 0.1 K in the boiling-point temperature causes a change of about 1% in the critical density for highly nonspherical molecules. Compared to experimental uncertainties of 1% to 3% and sometimes 5% in the present experimental data, this prediction is considered as very good.

Generally speaking, vapor pressures determined by the extended corresponding-states method lie within the experimental uncertainties for the most accurate data since the present method has been extensively tested. The uncertainty of the present method was determined to be approximately 0.05% to 0.1% in the vapor pressure for substances which have highly accurate experimental data, even in a limited range, which is comparable to the accuracy of the data. The uncertainty is also less than 0.1% for the first derivative and for the second derivative.

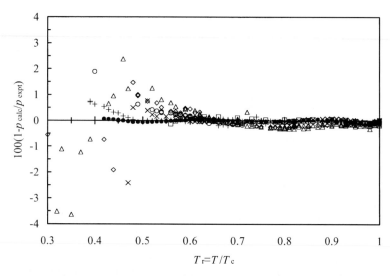

Fig. 6.4 Comparison of the highly accurate experimental data for substances which were used to determine the general coefficients for vapor pressure in the extended corresponding-states principle (□) Argon; (△) Ethane; (◊) Propane; (×) 1,1,1,2-tetrafluoroethane (R134a); (○) 1, 1- difluoroethane (R152a); (+) difluoromethane (R32); (●)Water. [Adapted from Xiang (2001a)]

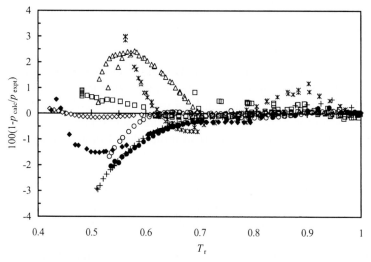

Fig. 6.5 Comparison of vapor pressure data for some highly polar substances with values calculated from the extended corresponding-states principl (◊) Water; (□) Ammonia; (△) Acetic acid; (*) Methanol; (○) Acetonitrile; (●) Diethyl ether; (+) Acetone; (♦) Decane. [Adapted from Xiang (2001a)]

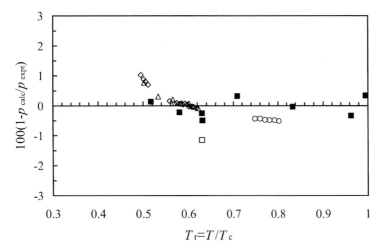

Fig. 6.6 Comparison of vapor-pressure data for CCl₃F with values calculated from the extended corresponding-states principle (■) Benning and McHarness (1940); (◊) Fernandez-Fassnacht and del Rio (1985); (Δ) Osborne et al. (1941); (○) Yurttas et al. (1990); (□) Varshni and Mohan (1954). [Adapted from Xiang (2001b)]

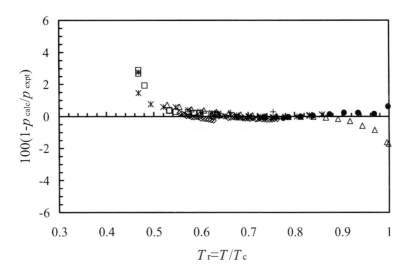

Fig. 6.7 Comparison of vapor-pressure data for CCl₂F₂ with values calculated from the extended corresponding-states principle (□) Blanke and Weiss (1992); (◊) Fernandez-Fassnacht and del Rio (1985); (Δ) Gilkey et al. (1931); (♦) Gordon and MacWilliam (1946); (*) Haendel et al. (1992); (●) Michels et al. (1966); (+) Mollerup and Fredenslund (1976). [Adapted from Xiang (2001b)]

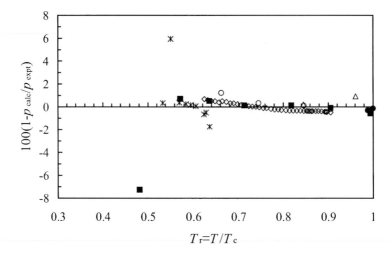

Fig. 6.8 Comparison of vapor-pressure data for $CClF_3$ with values calculated from the extended corresponding-states principle (■) Albright and Martin (1952); (◊) Fernandez-Fassnacht and del Rio (1985); (Δ) Mollerup and Fredenslund (1976); (○) Stein and Proust (1971); (*) Thorton et al. (1933); (●) Weber (1989a). [Adapted from Xiang (2001b)]

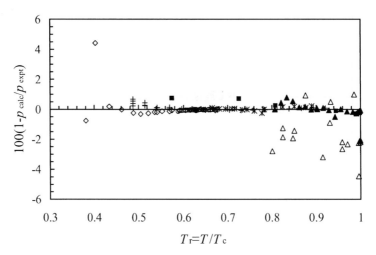

Fig. 6.9 Comparison of vapor-pressure data for $CHClF_2$ with values calculated from the extended corresponding-states principle (■) Benning and McHarness (1940); (◊) Blanke and Weiss (1992); (Δ) Booth and Swinehart (1935); (*) Giuliani et al. (1995); (○) Goodwin et al. (1992); (+) Haendel et al. (1992); (▲) Hongo et al. (1990). [Adapted from Xiang (2001b)]

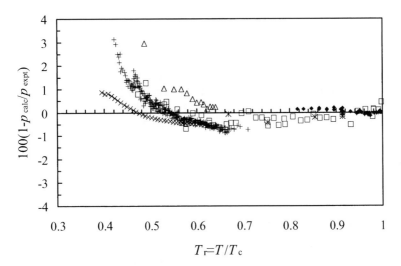

Fig. 6.10 Comparison of vapor-pressure data for CHF$_3$ with values calculated from the extended corresponding-states principle (♦) Hori et al. (1981); (□) Hou and Martin (1959); (×) Magee and Duarte-Garza (2000); (+) Popowicz et al. (1982); (*) Stein and Proust (1971); (Δ) Valentine et al. (1962). [Adapted from Xiang (2001b)]

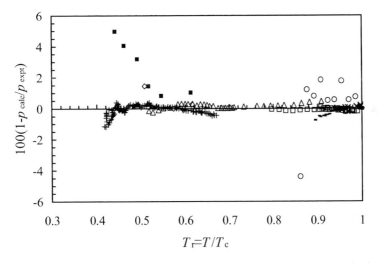

Fig. 6.11 Comparison of vapor-pressure data for CH$_3$F with values calculated from the extended corresponding-states principle (-) Biswas et al. (1989); (×) Bominaar et al. (1987); (○) Collie (1889); (□) Demiriz et al. (1993); (◊) Fonseca and Lobo (1994); (■) Grosse et al. (1940); (Δ) Michels and Wassenaar (1948); (+) Oi et al. (1983). [Adapted from Xiang (2001b)]

References

Albright, L. F. and J. J. Martin, 1952, *Ind. Eng. Chem.* **44**: 188.

Ambrose, D., 1978, *J. Chem. Thermodyn.* **10**: 765.

Ambrose, D., 1981, *J. Chem. Thermodyn.* **13**: 1161.

Ambrose, D. and I. J. Lawrenson, 1972, *Process Tech. Int.* **17**: 968.

Ambrose, D. and N. C. Patel, 1984, *J. Chem. Thermodyn.* **16**: 459.

Ambrose, D. and R. Townsend, 1963, *J. Chem. Soc.* 3614.

Ambrose, D., B. E. Broderick, and R. Townsend, 1967, *J. Chem. Soc. A* 633.

Ambrose, D., C. H. S. Sprake, and R. Townsend, 1974, *J. Chem. Thermodyn.* **6**: 693.

Ambrose, D., C. H. S. Sprake, and R. Townsend, 1975, *J. Chem. Thermodyn.* **7**: 185.

Ambrose, D., J. F. Counsell, and A. J. Davenport, 1970, *J. Chem. Thermodyn.* **2**: 283.

Ambrose, D., E. B. Ellender, D. A. Lee, C. H. S. Sprake, and R. Townsend, 1975, *J. Chem. Thermodyn.* **7**: 453.

Ambrose, D., J. H. Ellender, C. H. S. Sprake, and R. Townsend, 1977, *J. Chem. Thermodyn.* **9**: 735.

Ambrose, D., J. H. Ellender, H. A. Gundry, D. A. Lee, and R. Townsend, 1981, *J. Chem. Thermodyn.* **13**: 795.

Antoine, C., 1888, *Compt. Rend. Acad. Sci.* **107**: 681, 836, 1143.

Aston, J. G., C. W. Siller, and G. H. Messerly, 1937, *J. Am. Chem. Soc.* **59**: 1743.

Baehr, H. D. and R. Tillner-Roth, 1991, *J. Chem. Thermodyn.* **23**: 1063.

Baehr, H. D., H. Garnjost, and R. Pollak, 1976, *J. Chem. Thermodyn.* **8**: 113.

Benning, A. F. and R. C. McHarness, 1940, *Ind. Eng. Chem.* **32**: 497.

Biswas, S. N., C. A. Ten Seldam, S. A. R. C. Bominaar, and N. J. Trappeniers, 1989, *Fluid Phase Equil.* **49**: 1.

Blanke, W. and R. Weiss, 1992, PTB-Ber. W-48, Braunschweig.

Blanke, W. and R. Weiss, 1992a, *Fluid Phase Equil.* **80**: 179.

Blanke, W., G. Klingenberg, and R. Weiss, 1995, *Int. J. Thermophys.* **16**: 1143.

Bominaar, S. A. R. C., S. N. Biswas, N. J. Trappeniers, and C. A. Ten Seldam, 1987, *J. Chem. Thermodyn.* **19**: 959.

Booth, H. S. and C. F. Swinehart, 1935, *J. Am. Chem. Soc.* **57**: 1337.

Boublik, T., V. Fried, and E. Hala, 1984, The Vapour Pressures of Pure Substances, 2nd revised ed. (Elsevier, New York).

Boyes, S. J. and L. A. Weber, 1994, *Int. J. Thermophys.* **15**: 443.

Campbell, A. N. and R. M. Chatterjee, 1968, *Can. J. Chem.* **46**: 575.

Carruth, G. F. and R. Kobayashi, 1973, *J. Chem. Eng. Data* **18**: 115.

Chase, J. D., 1984, *Chem. Eng. Prog.* **80**: 63.

Chirico, R. D., A. Nguyen, W. V. Steele, M. M. Strube, and C. Tsonopoulos, 1989, *J. Chem. Eng. Data* **34**: 149.

Collie, N., 1889, *J. Chem. Soc.* **55**: 110.

Connolly J. F. and G. A. Kandalic, 1962, *J. Chem. Eng. Data* **7**: 137.

Cooney, A. and K. W. Morcom, 1988, *J. Chem. Thermodyn.* **20**: 1469.

Cox, E. R., 1936, *End. Eng. Chem.* **28**: 613.

Cragoe, C. S., C. H. Meyers, and C. S. Taylor, 1920, *J. Am. Chem. Soc.* **42**: 206.

Defibaugh, D. R., G. Morrison, and L. A. Weber, 1994, *J. Chem. Eng. Data* **39**: 333.

Demiriz, A. M., R. Kohlen, C. Koopmann, D. Moeller, P. Sauermann, G. A. Iglesias-Silva, and F. Kohler, 1993, *Fluid Phase Equil.* **85**: 313.

De Vries, B., 1997, Thermodynamische Eigenschaften der Alternativen Kltemittel R-32, R-125, R-143a, DKV-Forsch.-Ber. Nr. 55 (DKV, Stuttgart).

Duschek, W., R. Kleinrahm, and W. Wagner, 1990, *J. Chem. Thermodyn.* **22**: 841.

Edalat, M., R. B. Bozar-Jomehri, and G. A. Mansoori, 1993, *Oil Gas J.* **91(5)**: 39.

Fernandez-Fassnacht, E. and F. del Rio, 1985, *Cryogenics* **25**: 204.

Flebbe, J. L., D. A. Barclay, and D. B. Manley, 1982, *J. Chem. Eng. Data* **27**: 405.

Fonseca, I. M. A. and L. Q. Lobo, 1994, *J. Chem. Thermodyn.* **26**: 671.

Frost, A. A. and D. R. Kalkwarf, 1953, *J. Chem. Phys.* **21**: 264.

Fu, Y. D., L. Z. Han, and M. S. Zhu, 1995, *Fluid Phase Equil.* **111**: 273.

Gibbard, H. F. and J. L. Creek, 1974, *J. Chem. Eng. Data* **19**: 308.

Gilgen, R., R. Kleinrahm, and W. Wagner, 1994, *J. Chem. Thermodyn.* **26**: 399.

Gilkey, W. K., F. W. Gerard. and M. E. Bixler, 1931, *Ind. Eng. Chem.* **23**: 364.

Giuliani, G., S. Kumar, and F. Polonara, 1995, *Fluid Phase Equil.* **109**: 265.

Goodwin, R. D., 1969, *J. Res. Natl. Bur. Stand.* **73A**: 487.

Goodwin, A. R. H., D. R. Defibaugh, G. Morrison, and L. A. Weber, 1992a, *Int. J. Thermophys.* **13**: 999.

Goodwin, A. R. H., D. R. Defibaugh, and L. A. Weber, 1992, *Int. J. Thermophys.* **13**: 837.

Gordon, A. R. and E. A. MacWilliam, 1946, *Can. J. Res.* **24**: 292.

Grosse, A. V., R. C. Wackher, and C. B. Linn, 1940, *J. Phys. Chem.* **44**: 275.

Gupta, P. A. and D. E. Daubert, 1985, *Ind. Eng. Chem. Process Des. Dev.* **24**: 674.

Haendel, G., R. Kleingahm, and W. Wagner, 1992, *J. Chem. Thermodyn.* **24**: 697.

Haggenmacher, J. E., 1946, *J. Am. Chem. Soc.* **68**: 1633.

Hongo, M., M. Kusunoki, H. Matsuyama, T. Takagi, K. Mishima, and Y. Arai, 1990, *J. Chem. Eng. Data* **35**: 414.

Hori, K., S. Okazaki, M. Uematsu, and K. Watanabe, 1982, An Experimental Study of Thermodynamic Properties of Trifluoromethane, in Proc. 8th Symp. Thermophys. Prop. Vol. 2, J. V. Sengers ed. (ASME, New York).

Hou, Y. C. and J. J. Martin, 1959, *AIChE J.* **5**: 125.

Hsu, C. C. and J. J. McKetta, 1964, *J. Chem. Eng. Data* **9**: 45.

IAPWS, 1994, Skeleton Tables 1985 for the Thermodynamic Properties of Ordinary Water Substance, Int. Assoc. Prop. Water and Steam.

Iglesias-Silva, G. A., J. C. Holste, P. T. Eubank, K. N. Marsh, and K. R. Hall, 1987, *AIChE J.* **33**: 1550.

Kemp, J. D. and C. J. Egan, 1938, *J. Am. Chem. Soc.* **60**: 1521.

Kennedy, R. M., M. Sagenkahn, and J. G. Aston, 1941, *J. Am. Chem. Soc.* **63**:

2267.

Kleinrahm, R. and W. Wagner, 1986, *J. Chem. Thermodyn.* **18**: 739.

Kratzke, H. and S. Mueller, 1984, *J. Chem. Thermodyn.* **16**: 1157.

Kratzke, H. and S. Mueller, 1985, *J. Chem. Thermodyn.* **17**: 151.

Kretschmer, C. B. and R. Wiebe, 1949, *J. Am. Chem. Soc.* **71**: 1793.

Kretschmer, C. B. and R. Wiebe, 1952, *J. Am. Chem. Soc.* **74**: 1276.

Lee, B. I. and M. G. Kesler, 1975, *AIChE J.* **21**: 510.

Levelt Sengers, J. M. H., B. Kamgar-Parsi, F. W. Balfour and J. V. Sengers, 1983, *J. Phys. Chem. Ref. Data* **12**: 1.

Li, J. C. M. and K. S. Pitzer, 1956, *J. Am. Chem. Soc.* **78**: 1077.

Lydersen, A. L. and V. Tsochev, 1990, *Chem. Eng. Tech.* **13**: 125.

Magee, J. W., 1996, *Int. J. Thermophys.* **17**: 803.

Magee, J. W. and J. B. Howley, 1992, *Int. J. Refrig.* **15**: 362.

Magee, J. W. and H. A. Duarte-Garza, 2000, *Int. J. Thermophys.* **21**: 1351.

Malbrunot, P. F., P. A. Meunier, G. M. Scatena, W. H. Mears, K. P. Murphy, and J. V. Sinka, 1968, *J. Chem. Eng. Data* **13**: 16.

Mansoorian, H., K. R. Hall, J. C. Holste, and P. T. Eubank, 1981, *J. Chem. Thermodyn.* **13**: 1001.

McGarry, J., 1983, *Ind. Eng. Chem. Process, Des. Dev.* **22**: 313.

Messerly, G. H. and J. G. Aston, 1940, *J. Am. Chem. Soc.* **62**: 886.

Michels, A. and T. Wassenaar, 1948, *Physica* **14**: 104.

Michels, A., T. Wassenaar, G. J. Wolkers, C. Prins, and L. Klundert, 1966, *J. Chem. Eng. Data* **11**: 449.

Miller, D. G., 1964, *J. Phys. Chem.* **68**: 1399.

Moldover, M. R., 1985, *Phys. Rev. A* **31**: 1022.

Moldover, M. R. and J. C. Rainwater, 1988, *J. Chem. Phys.* **88**: 7772.

Mollerup, J. and A. Fredenslund, 1976, *J. Chem. Eng. Data* **21**: 299.

Mousa, A. H. N., 1981, *J. Chem. Thermodyn.* **13**: 201.

Nowak, P., R. Kleinrahm, and W. Wagner, 1996, *J. Chem. Thermodyn.* **28**: 1441.

Nowak, P., R. Kleinrahm, and W. Wagner, 1997, *J. Chem. Thermodyn.* **29**: 1157.

Oi, T., J. Shulman, A. Popowicz, and T. Ishida, 1983, *J. Phys. Chem.* **87**: 3153.

Osborne, D. W., C. S. Garner, R. N. Doescher, and D. M. Yost, 1941, *J. Am. Chem. Soc.* **63**: 3496.

Osborne, D. W., R. N. Doescher, and D. M. Yost, 1942, *J. Am. Chem. Soc.* **64**: 169.

Pal, A. K., G. A. Pope, Y. Arai, N. F. Carnaha, and R. Kobayashi, 1976, *J. Chem. Eng. Data* **21**: 394.

Popowicz, A., T. Oi, J. Shulman, and T. Ishida, 1982, *J. Chem. Phys.* **76**: 3732.

Potter, A. E. and H. L. Ritter, 1954, *J. Phys Chem.* **58**: 1040.

Rainwater, J. C. and J. J. Lynch, 1989, *Fluid Phase Equil.* **52**: 91.

Reid, R.C. and T. K. Sherwood, 1958, The Properties of Gases and Liquids (McGraw-Hill, New York).

Reid, R.C. and T. K. Sherwood, 1966, The Properties of Gases and Liquids, 2nd ed. (McGraw-Hill, New York).

Reid, R.C., J. M. Prausnitz, and T. K. Sherwood, 1977, The Properties of Gases and Liquids, 3rd ed. (McGraw-Hill, New York).

Reid, R. C., J. M. Prausnitz, and B. E. Poling, 1987, The Properties of Gases and Liquids, 4th ed. (McGraw-Hill, New York).

Riedel, L., 1954, *Chem. Ing. Tech.* **26**: 83.

Ruzicka, K. and V. Majer, 1994, *J. Phys. Chem. Ref. Data* **23**: 1.

Sato, T., H. Sato, and K. Watanabe, 1994, *J. Chem. Eng. Data* **39**: 851.

Scott, D. W. and A. G. Osborn, 1979, *J. Phys. Chem.* **83**: 2714.

Silva, A. M. and L. A. Weber, 1993, *J. Chem. Eng. Data* **38**: 644.

Smith, B. D. and R. Srivastava, 1986, Thermodynamic Data for Pure Compounds (Elsevier, Amsterdam).

Spence, R. and W. Wild, 1935, *J. Chem. Soc.* 506.

Stein, F. P. and P. C. Proust, 1971, *J. Chem. Eng. Data* **16**: 389.

Straty, G. C. and R. Tsumura, 1976, *J. Res. Natl. Bur. Stand.* **80A**: 35.

Stull, D. R., 1947, *Ind. Eng. Chem.* **39**: 517.

Taylor, R. S. and L. B. Smith, 1922, *J. Am. Chem. Soc.* **44**: 2450.

Thek, R. E. and L. I. Stiel, 1966, *AIChE J.* **12**: 599.

Thomas, L. H., 1976, *Chem. Eng. J.* **11**: 191.

Thomas, R. H. P. and R. H. Harrison, 1982, *J. Chem. Eng. Data* **27**: 1.

Thornton, N. V., A. B. Burg, and H. I. Schlesinger, 1933, *J. Am. Chem. Soc.* **55**: 3177.

Tsvetkov, O. B., A. V. Kletski, Yu. A. Laptev, A. J. Asambaev, and I. A. Zausaev, 1995, *Int. J. Thermophys.* **16**: 1185.

Valentine, R. H., G. E. Brodale, and W. F. Giauque, 1962, *J. Phys. Chem.* **66**: 392.

Varshni, Y. P. and H. Mohan, 1954, *J. Indian Chem. Soc.* **32**: 211.

Vetere, A., 1986, *Chem. Eng. J.* **32**: 77.

Wagner, W., 1973, *Cryogenics* **13**: 470.

Waring, W., 1954, *Ind. Eng. Chem.* **46**: 762.

Watson, K. M., 1948, *Ind. Eng. Chem.* **35**: 398.

Weber, L. A., 1989, *Int. J. Thermophys.* **10**: 617.

Weber, L. A. and A. R. H. Goodwin, 1993, *J. Chem. Eng. Data* **38**: 254.

Weber, L. A. and A. M. Silva, 1994, *J. Chem. Eng. Data* **39**: 808.

Wilding, W. V., J. K. Johnson, and R. L. Rowley, 1987, *Int. J. Thermophys.* **8**: 717.

Wilding, W. V. and R. L. Rowley, 1986, *Int. J. Thermophys.* **7**: 525.

Wilson, D. P. and R. S. Basu, 1988, *ASHRAE Trans.* **94**: 2095.

Wu, G. Z. A. and L. I. Stiel, 1985, *AIChE J.* **31**: 1632.

Xiang, H. W., 1995, New Vapor Pressure Equation and Crossover Equation of State, Doctoral Dissertation, Jiaotong Univ., Xi'an, China.

Xiang, H. W., 2001a, *Int. J. Thermophys.* **22**: 919.

Xiang, H. W., 2001b, *J. Phys. Chem. Ref. Data* **30**: 1161.

Xiang, H. W., 2002, *Chem. Eng. Sci.* **57**: 1439.

Xiang, H. W., 2003, Thermophysicochemical Properties of Fluids: Corresponding -States Principle and Practise (Science Press, Beijing), in Chinese.

Xiang, H. W. and L. C. Tan, 1994, *Int. J. Thermophys.* **15**: 711.

Xu, Z., 1984, *Ind. Eng. Chem. Process Des. Dev.* **23**: 7.

Yurttas, L., J. C. Holste, K. R. Hall, B. E. Gammon, and K. N. Marsh, 1990, *Fluid Phase Equil.* **59**: 217.

Zander, M. and W. Thomas, 1979, *J. Chem. Eng. Data* **24**: 1.

Chapter 7

Transport Properties

7.1 Introduction

Accurate knowledge of transport properties of gases is essential for the optimum design of the different items of chemical processing plants, for determination of intermolecular potential functions and for development of accurate theories of transport properties in dense fluid states. They are especially of interest because a framework for an understanding of intermolecular forces can be provided. It is impossible to measure these properties for all industrially and theoretically important fluids; therefore, it is very necessary to predict these properties by use of suitable theoretical models.

The transport properties in a dilute gas or vapor phase can be predicted from the kinetic theory. Existing methods are limited to a specific class of fluids and can not be applied to various substances. The Enskog theory exists for the transport properties of hard-sphere dense liquids. No model can be found in the reviews of Chapman and Cowling (1939, 1952, 1970), Reid et al. (1977, 1987), and Chung et al. (1984) to predict the transport properties over the entire region consistently from the dilute gas to the liquid state.

Of the many predictive approaches that have been proposed, the corresponding-states principle has proved to be the most powerful framework (Leland and Chappelear, 1968; Huber and Hanley, 1996; Xiang, 2003). The corresponding-states principle was the most useful and accurate derivation of the Van der Waals equation of state and has a firm basis in statistical mechanics and kinetic theory, and has great range and accuracy. The methods based on the corresponding-states principle are theoretically based and predictive, rather than empirical and correlative. While the corresponding-states principle cannot always reproduce a set of data within its experimental accuracy, as can an empirical correlation, nevertheless it should not only be able to represent data to a reasonable degree but, more importantly it does what a correlation cannot do, that is, predict the properties beyond the range of existing data.

The corresponding-states theory of Pitzer et al. has been widely used, as it characterizes the behavior of normal molecular substances and proved to be successful for many nonpolar small molecular systems; however, its predictive capability for polar and associating molecules is generally poor. The extended corresponding-states theory was established based on the aspherical factor to successfully extend the corresponding states theory of Pitzer et al. to systems containing highly polar species and species exhibiting specific interactions like associating and hydrogen bonding. As an improvement on the corresponding-states theory of Pitzer et al., the extended corresponding-states theory may accurately describe and predict the transport properties of polar and associating molecules.

The theory of gas transport properties has been reasonably well clarified by the application of the kinetic theory of gases; however, the theory of liquid viscosity is poorly developed. This chapter will briefly present both theories, introduce the extended corresponding states method to describe the transport properties of various real substances, and predict the transport properties over the entire region of different classes of substances with the Enskog hard-sphere dense theory. In this chapter, the theory of gas transport properties is first presented, the theoretical basis and calculated method for transport properties in the critical region is then discussed, and finally the transport properties of liquids is correlated.

7.2 Theory of Gas Transport Properties

It should be noted that the transport properties differ in one important respect from the properties discussed previously in this book; namely, transport properties are nonequilibrium on a macroscale. Density, for example, is an equilibrium property. On a microscale, both properties reflect the effect of molecular motion and interaction. Even though the transport properties are ordinarily referred to as nonequilibrium properties, they are, like temperature, pressure, and volume, a function of the state of the fluid, and may be used to define the state of the material. Brule and Starling (1984) emphasized the desirability of using both transport and thermodynamic data to characterize complex fluids and to develop correlations.

If a shearing stress is applied to any portion of a confined fluid, the fluid will move and a velocity gradient will be set up within it, with a maximum velocity at the point where the stress is applied. If the shear stress per unit area at any point is divided by the velocity gradient, the ratio obtained is defined as the viscosity of the medium. It can be seen, therefore, that viscosity is a measure of the internal fluid friction, which tends to oppose any dynamic change in the fluid motion; i.e., if the friction between layers of fluid is small, an applied shearing force will result in a large velocity gradient. As the viscosity increases, each fluid layer exerts a larger frictional drag on adjacent layers and the velocity gradient decreases.

The theory of gas transport properties is simply stated; however, gas transport properties can only be expressed in quite complex equations that can be used directly to calculate transport properties. In simple terms, when a gas undergoes a shearing stress so that there is some bulk motion, the molecules at any one point have the bulk velocity vector added to their own random velocity vector.

Molecular collisions cause an interchange of momentum throughout the fluid, and this bulk motion velocity or momentum becomes distributed. Near the source of the applied stress, the bulk velocity vector is high, but as the molecules move away from the source, they are slowed down in the direction of bulk flow, which causes the other sections of the fluid to move in that direction. This random, molecular momentum interchange is the predominant cause of gaseous viscosity.

For molecules that attract or repel one another by·virtue of intermolecular forces, the theory of Chapman and Enskog (Chapman and Cowling, 1939; Hirschfelder et al., 1964) is normally employed. There are four important assumptions in this theory: (1) the gas is sufficiently dilute for only binary collisions to occur, (2) the motion of the molecules during a collision can be described by classical mechanics, (3) only elastic collisions occur, and (4) the intermolecular forces act only between fixed centers of the molecules; i. e., the intermolecular potential function is spherically symmetric. With these restrictions, it would appear that the resulting theory should be applicable only to low-pressure, high-temperature monatomic gases. The pressure and temperature restrictions are valid, but due to a lack of tractable, alternate models, it is very often applied to polyatomic gases except in the case of thermal conductivity, for which a correction for internal energy transfer and storage must be included.

The Chapman-Enskog theory treats the interactions between colliding molecules in detail, with the potential energy included. The equations are well known, but their solution is often very difficult. Each choice of an intermolecular potential must be solved separately.

7.2.1 Corresponding States of Zero-Density Gas Viscosity

The theoretical models for gas or vapor-phase viscosity are based on the kinetic theory of gases. According to the rigorous kinetic theory of gases, the Chapman-Enskog (Chapman and Cowling, 1952, 1970; Hirschfelder et al., 1964) theory for the dilute gas viscosity η_{CE} is given by

$$\eta_{CE} = 5(mkT)^{1/2} / 16\pi^{1/2}\sigma^2\Omega^{*(2,2)} \tag{7.1}$$

where k is the Boltzmann constant, T is the temperature, and $m = M/N_A$ is the molecular mass, with M being the molar mass and N_A the Avogadro number. The collision diameter σ is defined as the separation distance when the intermolecular potential function is equal to zero. The reduced collision integral $\Omega^{*(2,2)}$ was determined from the Lennard-Jones intermolecular potential and is a function of the reduced temperature, $T^* = kT/\varepsilon$, where ε is the potential energy of interaction between the molecules. For hard spheres, the collision integral $\Omega^{*(2,2)}$ is equal to unity; as a result, η_{CE} approaches the dilute gas viscosity η_0. The reduced collision integral, $\Omega^{*(2,2)}$, was correlated by Neufeld et al. (1972) as follows:

$$\Omega^{*(2,2)} = 1.16145(T^*)^{-0.14874} + 0.52487\exp(-0.77320T^*)$$
$$+ 2.16178\exp(-2.43787T^*) \tag{7.2}$$

Eq. (7.2) is a universal expression for the viscosity of the Lennard-Jones potential in the zero-density limit and is applicable from 0.3 to 100 in T^*, with an average deviation of 0.064% (Neufeld et al., 1972). When the Lennard-Jones parameters ε and σ are available, the gas viscosity can be calculated from Eqs. (7.2) and (7.1). However, the parameters ε and σ are usually determined from experimental viscosity data in the zero-density limit. The reported values of the Lennard-Jones potential parameters ε and σ often differ. Nevertheless, it is should be noted that, between reported sets of potential parameters, usually if a large positive difference exists in, say, ε, then there is a large negative difference in the reported value of σ. This fact often causes the calculated gas viscosity to be almost the same, as there is a compensation effect. According to Chung et al. (1984), the two parameters ε and σ are evaluated from the critical density and critical temperature for all substances using the following empirical relations:

$$\varepsilon / k = gT_c \quad \text{and} \quad \sigma\rho_c^{1/3} = h, \tag{7.3}$$

where T_c and ρ_c are the critical density and critical temperature, respectively. Lennard-Jones potential parameters to accurately describe the viscosity in the zero-density limit for argon are $\sigma = 0.335$ nm and $\varepsilon/k = 143$ K (Aziz and Slaman, 1986; Bich et al., 1990). As a result, the corresponding values of h and g for argon in Eq.(7.3) are 0.669 and 0.93, respectively; and $h = 0.669$ is used for any molecule. According to the extended corresponding-states theory of complex molecules, g, which is required later in this work, can be correlated as follows:

$$g = 0.93 + u_1\omega + u_2\theta, \tag{7.4}$$

where $\omega = -1 - \log p_r|_{T_r=0.7}$ is the acentric factor, $\theta = (Z_c - 0.29)^2$ is the aspherical factor, p_r is the reduced vapor pressure, and $T_r = T/T_c$ is the reduced temperature. The values of u_1 and u_2 are given in Table 7.1.

In the reduced form, Eq. (7.1) can be written as follows:

$$\eta^*_{CE} = T^{*1/2} / \Omega^{*(2,2)}(T^*) . \tag{7.5}$$

The reduced viscosity η^* (Chung et al., 1984) is defined as:

$$\eta^* = \left[5(mkT/\pi)^{1/2} / 16\sigma^2\right]^{-1} \eta . \tag{7.6}$$

To account for the anisotropic effects, f, as a correction factor, was introduced to make the Chapman-Enskog theory suitable for prediction of the viscosity of nonspherical gases. The resulting equation is given as (Chung et al., 1984):

$$\eta^*_0 = \eta^*_{CE} f . \tag{7.7}$$

where η_0 is the dilute gas viscosity, and $f = 1$ for spherical molecules. According to the extended corresponding-states theory,

$$f = u_0 + u_1\omega + u_2\theta . \tag{7.8}$$

The values of u_0, u_1 and u_2 are given in Table 7.1.

7.2.2 Corresponding States of Zero-Density Gas Thermal Conductivity

For the thermal conductivity λ, Chung et al. (1984) derived the following expression based on the approximations of Mason and Monchick (1962) from the Wang Chang-Uhlenbeck (1951) and from the Taxman (1958) kinetic theories for polyatomic gases,

$$\lambda^*_0 / \eta^*_0 = 1 + c^*_{int}\left[\frac{0.26665 + (0.215 - 1.061\beta + 0.28288c^*_{int})/Z_{coll}}{\beta + (0.6366 + 1.061\beta c^*_{int})/Z_{coll}}\right], \tag{7.9}$$

where c^*_{int} is the internal heat capacity at constant volume, $\beta = \eta/\rho D$ is defined as the ratio of the viscosity η and the product of density ρ and internal self-diffusivity D in the dilute gas state. The reduced thermal conductivity is defined as (Chung et al., 1984)

$$\lambda^* = \left[75k/64(m\pi/\varepsilon)^{1/2}\sigma^{-2}\right]^{-1}\lambda . \tag{7.10}$$

The key to its use is the accurate estimation of the rotational collision number

Z_{coll}. To date, this is still not possible, although many authors have discussed the problem, as stated by Reid et al. (1987). Z_{coll} is the collision number required to interchange a quantum of internal energy with translational energy which can be correlated as a function of temperature as:

$$Z_{coll} = 2 + (c_0 + c_1\omega + c_2\theta)(kT/\varepsilon)^2, \qquad (7.11)$$

where the parameters c_0, c_1, and c_2 are given in Table 7.1. It should be noted that the first term in Eq. (7.11) is empirical (Chung et al., 1984). The reduced internal heat capacity at constant volume is expressed as $c_{int}^* = c_{int}/R$, with c_{int} being the internal heat capacity and R the gas constant; i.e., $c_{int}^* = c_V^* - c_{tr}^*$ and $c_{tr}^* = 3/2$, where c_V^* can be obtained from any available correlation for $c_V^* = c_P^* - 1$. However, it is difficult to obtain the value of β through this definition, because of the lack of experimental data for the internal self-diffusivity (Chung et al. 1984). To do this, Chung et al. (1984) treated β as a parameter, which can only be determined using a reliable method. β was determined from the extended corresponding-states theory:

$$\beta = l_0 + l_1\omega + l_2\theta. \qquad (7.12)$$

The values of l_0, l_1, and l_2 are given in Table 7.1. A comparison of the calculated results for β in Eq. (7.9) showed that it was in a very good agreement with the results obtained from the substance-dependent parameters given by Chung et al. (1984) which describes the recommended thermal conductivity data within 5 % generally.

As a result, f in Eq. (7.8) for viscosity and β in Eq. (7.12) for thermal conductivity were introduced to extend the Chapman-Enskog theory to make it suitable for polyatomic, polar, hydrogen-bonding, and associating gases in a generally consistent way for any real molecule.

The general parameters of Eqs. (7.4), (7.8), (7.11), and (7.12) were determined by fitting experimental and recommended viscosity and thermal conductivity data for methane, ethane, carbon dioxide, propane, 1, 1, 1, 2-tetrafluoroethane, 1,1-difluoroethane, ammonia, and water as simple, nonpolar and polar substances. These substances along with their critical parameters, acentric factors, aspherical factors and data sources are listed in Tables 7.2 and 7.3. It should be noted that these data usually contain deviations of a few percentage, generally 5% or slightly higher. The general parameters given in Table 7.1 were independent of the temperature, density, and specific substance and were universal for all classes of molecules. As shown in Figs. 7.1 and 7.2, the representation of the dilute gas viscosity data is reasonably within the experimental uncertainties for nonpolar and polar substances. In the figures, only one symbol is used for all of the data for each molecule for the sake of clarity. The present results for dilute vapor viscosity were also in good agreement with a usual temperature range, at the reduced temperature of about 0.7, for halogenated, hydrogen-bonding and associating substances, such as chlorodifluoromethane, difluoromethane, 1, 1-dichloro-2, 2, 2-trifluoroethane, 1, 1, 1, 2-tetrafluoroethane, 1, 1-difluoroethane, water, methanol, ammonia, and acetone. All of the calculated values agree reasonably well, usually within 5% and sometimes slightly larger than 10%.

As shown in Figs. 7.3 and 7.4, the representation of the thermal conductivity was reasonably good, within the possible uncertainties in the data for nonpolar and polar substances. It should be noted that the data accuracy is generally a few percentages. One can see in Fig. 7.3 that the differences between the two recommended data sets obtained by Uribe et al. (1990) and Laesecke et al. (1990) are 5-10% even for oxygen, such a simple molecule, in the low and high temperature ranges. Generally, data sets have an agreement of 5% and sometimes 10%, although the claimed accuracy is higher. The present corresponding-states theory gave a good description in all classes of molecules in contrast to the existing models applicable to a limited range of molecules (Chapman and Cowling, 1970; Reid et al., 1977, 1987; Chung et al., 1984).

The extended corresponding-states theory was used to describe and predict the transport properties of real molecules including highly polar and hydrogen-bonding molecules based on the Chapman-Enskog theory for dilute gases. The prediction of the transport properties from the present model is more general, reliable, and consistent with the existing methods, while only the critical parameters, acentric factor, and aspherical factor are required in the general scheme. Considering the fact that transport properties are usually less studied theoretically, this work improves the theory in order to gain a better understanding of intermolecular forces and to examine the reliability of experimental data. This work demonstrates a close fit with the available predictive methods and predicts a general scheme for a wide set of substances within a simple and reliable corresponding-states framework. In contrast to the previous methods, the method presented here predicts the transport properties of nonpolar, polar, and associating substances in the dilute gas state and for high-density gases to be presented later.

Table 7.1 General Coefficients of Eqs. (7.4), (7.8), (7.11), and (7.12)

u_0	1	c_0	6	l_0	0.75
u_1	-0.25	c_1	-6	l_1	-0.65
u_2	50	c_2	600	l_2	125

Table 7.2 Molar Mass M, Critical Temperature T_c, Critical Pressure p_c, Critical Density ρ_c, Acentric Factor ω, and Aspherical Factor θ

Substance	M (kg/kmol)	T_c (K)	p_c (kPa)	ρ_c (kg.m^{-3})	ω	$\theta \times 10^3$
Argon	39.948	150.69	4863	535	0.000	0.000
Oxygen	31.999	154.6	5040	435	0.025	0.002
Nitrogen	28.013	126.19	3395	313	0.037	0.000
Sulfur hexafluoride	146.06	318.7	3750	742	0.220	0.131
Carbon monoxide	28.010	132.9	3500	300	0.066	0.033
Carbon dioxide	44.010	304.13	7377	468	0.225	0.245
Methane	16.043	190.56	4598	162	0.011	0.007
Ethane	30.070	305.32	4872	206	0.099	0.097
Propane	44.097	369.83	4248	220	0.152	0.171
n-Butane	58.123	425.12	3796	228	0.199	0.263
n-Pentane	72.150	469.7	3370	232	0.251	0.468
n-Hexane	86.177	507.6	3025	234	0.299	0.678
n-Heptane	100.204	540.2	2740	234	0.350	0.827
n-Octane	114.232	568.8	2490	234	0.398	1.087
Isobutane	58.124	408.2	3650	221	0.183	0.051
Neopentane	72.151	433.8	3196	235	0.197	0.322
Isopentane	72.151	460.4	3390	235	0.227	0.328
Cyclohexane	84.161	553.8	4080	273	0.212	0.284
Ethylene	28.054	282.35	5041	215	0.086	0.096
Propylene	42.081	364.9	4600	232	0.144	0.225
Acetylene	26.038	308.3	6140	231	0.190	0.400
Benzene	78.114	562.05	4895	305	0.210	0.472
Toluene	92.140	591.75	4108	292	0.263	0.704
p-Xylene	106.167	616.2	3511	281	0.320	0.966
Chlorodifluoro-methane	86.470	369.28	4988	525	0.221	0.501
Difluoromethane	52.020	351.26	5780	430	0.277	2.560
2,2-Dichloro-1,1,1-trifluoroethane	152.93	456.83	3662	550	0.282	0.481
1,1,1,2-Tetrafluoro-ethane	102.03	374.18	4055	511	0.327	0.885
1,1-Difluoroethane	66.050	386.41	4516	369	0.275	1.474
Ammonia	17.031	405.37	11345	234	0.256	2.028
Acetone	58.080	508.1	4700	270	0.308	2.569
Ethyl acetate	88.107	523.2	3830	308	0.362	1.455
Ethyl ether	74.123	466.7	3640	258	0.281	0.420
Sulfur dioxide	64.063	430.8	7880	524	0.256	0.443

Phenol	94.113	694.2	5930	411	0.438	2.997
Water	18.015	647.10	22050	325	0.344	3.947
Methanol	32.042	512.0	8000	270	0.560	4.486
Ethanol	46.069	513.9	6140	276	0.644	2.514
1-Propanol	60.096	536.8	5170	274	0.623	1.292
Acetic acid	60.052	593.0	5786	330	0.450	5.844
Acetonitrile	41.053	545.4	4830	225	0.338	9.151

Table 7.3 Data Source of Zero-Density Gas Viscosity η_0 and Thermal Conductivity λ_0

Substance	Refs. (η_0)	Refs. (λ_0)
Argon	Bich et al., 1990; Kestin et al., 1984; Younglove and Hanley, 1986	Bich et al., 1990
Oxygen	Boushehri et al., 1987; Uribe et al., 1990; Laesecke et al., 1990	Boushehri et al., 1987; Uribe et al., 1990; Laesecke et al., 1990
Nitrogen	Younglove and Hanley, 1986	Younglove and Hanley, 1986
Methane	Boushehri et al., 1987	Boushehri et al., 1987
Ethane	Boushehri et al., 1987	Boushehri et al., 1987
Propane	Boushehri et al., 1987; Hendl et al., 1991	Boushehri et al., 1987; Hendl et al., 1991
Butane	Vogel, 1995	Vargeftik, 1993
Pentane	Reid et al., 1977,1987	Vargeftik, 1993
Hexane	Reid et al., 1977,1987	Vargeftik, 1993
Heptane	Reid et al., 1977,1987	Vargeftik, 1993
Octane	Reid et al., 1977,1987	Vargeftik, 1993
Isobutane	Reid et al., 1977,1987	Vargeftik, 1993
Isopentane	Reid et al., 1977,1987	Vargeftik, 1993
Carbon monoxide	Boushehri et al., 1987	Boushehri et al., 1987
Carbon dioxide	Boushehri et al., 1987	Boushehri et al., 1987
Ethylene	Boushehri et al., 1987	Boushehri et al., 1987
Propylene	Reid et al., 1977,1987	Vargeftik, 1993
Acetylene	Reid et al., 1977,1987	Vargeftik, 1993
Benzene	Vogel et al., 1986	Vargeftik, 1993
Chlorodifluoro-methane	Kestin et al., 1979	Vargeftik, 1993
Difluoromethane	Takahashi et al., 1995	Krauss et al., 1993
1,1-Dichloro-2,2,2-trifluoroethane	Nabizadeh and Mayinger, 1992; Dowdell and Matthews, 1993	Vargeftik, 1993

1,1,1,2-Tetrafluoro-ethane	Nabizadeh and Mayinger, 1992; Dowdell and Matthews, 1993; Wilhelm and Vogel, 1996	Gross and Song, 1996
1,1-Difluoroethane	Takahashi et al., 1987; Krauss et al., 1996	Krauss et al., 1996
Ammonia	Fenghour et al., 1995	Fenghour et al., 1995
Acetone	Reid et al., 1977,1987	Reid et al., 1977,1987
Ethyl acetate	Reid et al., 1977,1987; Vargeftik, 1975	Vargeftik, 1993
Ethyl ether	Reid et al., 1977,1987	Vargeftik, 1993
Sulfur dioxide	Reid et al., 1977,1987	Vargeftik, 1993
Water	Sengers and Watson, 1986	Sengers and Watson, 1986
Methanol	Vogel et al., 1986	Reid et al., 1977,1987
Ethanol	Reid et al., 1977,1987	Vargeftik, 1993
1-Propanol	Reid et al., 1977,1987	Vargeftik, 1993
Acetic acid	Reid et al., 1977,1987	Reid et al., 1977,1987
Acetonitrile	Reid et al., 1977,1987	Reid et al., 1977,1987

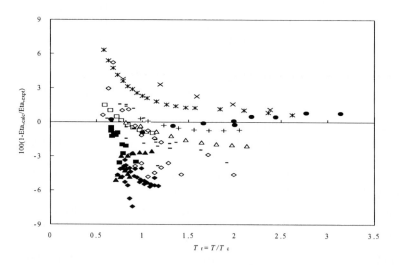

Fig. 7.1 Comparison of the viscosity in zero-density limit with the values calculated from the extended corresponding-states theory for argon (●), nitrogen (×), carbon dioxide (+), ethane (Δ), methane (*), benzene (□), chlorodifluoromethane (-), difluoromethane (◊), 1,1-dichloro-2, 2, 2-trifluoroethane (■), 1,1,1,2-tetrafluoroethane (♦), and 1,1-difluoroethane (▲). [Reproduced with permission from Xiang (2001)]

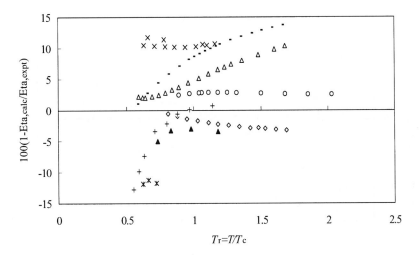

Fig. 7.2 Comparison of the viscosity in the zero-density limit with the values calculated from the extended corresponding-states theory for ethylene (○), propane (◊), acetone (▲), ammonia (Δ), water (-), methanol (×), acetic acid (+), and acetonitrile (*). [Reproduced with permission from Xiang (2001)]

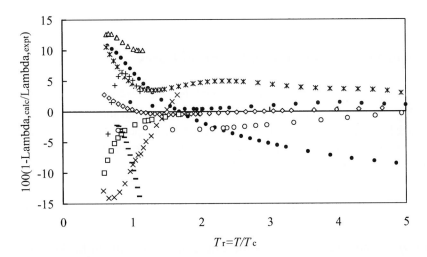

Fig. 7.3 Comparison of the thermal conductivity in the zero-density limit with the values calculated from the extended corresponding-states theory for oxygen (●), nitrogen (○), methane (*), argon (◊), 1,1,1,2-tetrafluoroethane (-), 1,1-difluoroethane (+), ammonia (□), water (×), and methanol (Δ). [Reproduced with permission from Xiang (2001)]

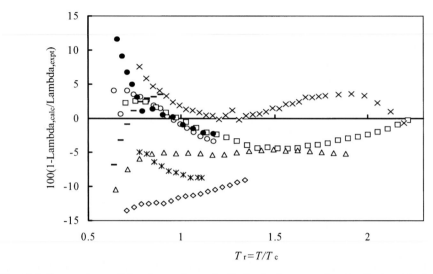

Fig. 7.4 Comparison of the thermal conductivity in the zero-density limit with the values calculated from the new extended corresponding-states principle for ethylene ($\times$), propane ($\square$), butane ($\triangle$), hexane ($\blacktriangle$), heptane ($\lozenge$), benzene ($\circ$), chlorodifluoromethane ($\bullet$), ethyl acetate (-), sulfur dioxide (+), and ethanol (*).[Reproduced with permission from Xiang (2001)]

7.2.3 Second Viscosity Virial Coefficients of Gases

A microscopically based theoretical model for the viscosity η and thermal conductivity λ of monatomic molecules was proposed that interacts according to the Lennard-Jones (12-6) potential for the second transport virial coefficients of gases (Rainwater, 1984; Friend and Rainwater, 1984; Rainwater and Friend, 1987). The second viscosity virial coefficient B_η is defined by

$$\eta = \eta_0\left(1 + B_\eta \rho + \cdots\right), \tag{7.13}$$

where η is the viscosity, η_0 is the dilute gas or Chapman-Enskog limit of the viscosity and ρ is the number density. The second thermal conductivity virial coefficient B_λ, where λ is thermal conductivity, may be defined similarly. For a Lennard-Jones 12-6 potential, their model agrees well for both transport virial coefficients with experimental results in the region $T^* \geq 1$, where $T^* = k_B T / \varepsilon$, k_B is Boltzmann's constant, and ε is the intermolecular potential energy parameter.

Vogel et al. (1986) determined B_η from their measurements of the viscosity of benzene and methanol vapors at $T^* < 1$, which show that B_η becomes negative and rapidly decreases with decreasing temperature. Rainwater and Friend (1987) presented a numerical result for the reduced transport virial coefficients as functions of T^* down to $0.5 \leq T^* \leq 0.9$. Comparison with the experimental viscosity data of Vogel et al. (1986) showed that the theoretical transport virial coefficients calculated with the Lennard-Jones potential agree with the results of experiments well within the limits that can be expected for polyatomic gases.

The reduced second viscosity virial coefficient B_η^* as a function of the reduced temperature T^* may be predicted from the extended corresponding-states theory and based on the Lennard-Jones 12-6 potential.

The second viscosity virial coefficient is represented as:

$$B_\eta^* = e_0 / T_r^{1/3} + e_1 / T_r + e_2 / T_r^2 + e_3 / T_r^3 + e_4 / T_r^6, \qquad (7.14)$$

where $B_\eta^* = B_\eta / \rho_c$, $T_r = T / T_c$ is the reduced temperature, with T being the temperature and T_c the critical temperature. The adjustable parameter e_i in Eq. (7.14) are depends on the substance and may be determined from the experimental data.

According to the extended corresponding-states theory, the corresponding-states theory describes the second viscosity virial coefficient as follows:

$$e_0 = e_{00}(1 + e_{01}\omega + e_{02}\theta)$$
$$e_1 = e_{10}(1 + e_{11}\omega + e_{12}\theta)$$
$$e_2 = e_{20}(1 + e_{21}\omega + e_{22}\theta)$$
$$e_3 = e_{30}(1 + e_{31}\omega + e_{32}\theta)$$
$$e_4 = e_{40}(1 + e_{41}\omega + e_{42}\theta) \qquad (7.15)$$

The general coefficients of Eq. (7.15) are given in Table 7.5. Figures 7.5 to 7.8 show how the extended corresponding-states method represents the second viscosity virial coefficient data for various molecules. In most cases, the agreement between the extended corresponding-states method and the experimental data listed in Table 7.4 is generally comparable to the agreement between the different sets of data. The extended corresponding-states theory provides a reliable method to determine the second viscosity virial coefficient for complex molecules.

Table 7.4 Data Source of the Second Viscosity Virial Coefficient B_η

Substance	Refs.
Argon	Rainwater and Friend (1987); Kestin et al. (1981); Kestin et al. (1971); Gracki et al. (1969)
Sulfur hexafluoride	Strehlow and Vogel (1989)
Ethane	Hendl and Vogel (1992)
Propane	Vogel (1995)
Butane	Kuchenmeister and Vogel (1998)
Isobutane	Kuchenmeister and Vogel (2000)
Neopentane	Vogel et al. (1988)
Cyclohexane	Vogel et al. (1988)
Hexane	Vogel and Strehlow (1988)
Benzene	Vogel et al. (1986)
Toluene	Vogel and Hendl (1992)
p-Xylene	Vogel and Hendl (1992)
1,1,1,2-Tetrafluoroethane	Wilhelm and Vogel (1996)
Phenol	Vogel and Neumann (1993)
Ammonia	Iwasaki and Takahashi (1968)
Sulfur dioxide	Iwasaki and Takahashi (1975)
Water	Alexandrov et al. (1975)
Methanol	Vogel et al. (1986)

Table 7.5 General Coefficients of Eq. (7.15)

e_{00}	−0.4755251	e_{01}	−7.054146	e_{02}	3153.843
e_{10}	2.420623	e_{11}	−4.502944	e_{12}	1487.502
e_{20}	−2.72150	e_{21}	−5.278855	e_{22}	1052.748
e_{30}	0.9306469	e_{31}	−7.745423	e_{32}	800.0538
e_{40}	−0.08942532	e_{41}	−6.275877	e_{42}	359.3725

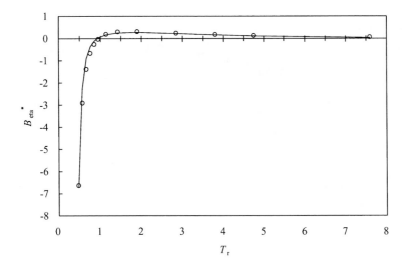

Fig. 7.5 Comparison of the second viscosity virial coefficient data with the values calculated from the extended corresponding-states theory for argon.

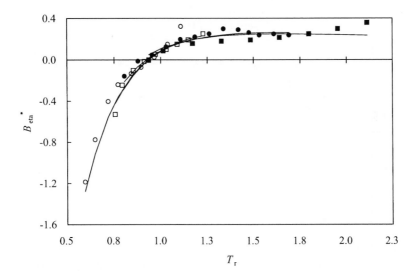

Fig. 7.6 Comparison of the second viscosity virial coefficient data with the values calculated from the extended corresponding-states theory for sulfur hexafluoride (■), propane (●), hexane (□), benzene (○).

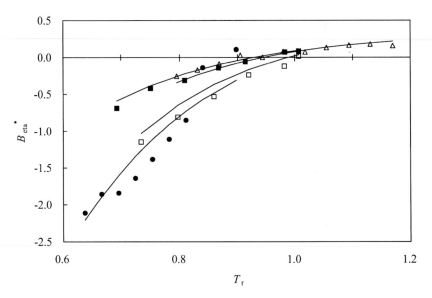

Fig. 7.7 Comparison of the second viscosity virial coefficient data with the values calculated from the extended corresponding-states theory for sulfur dioxide (■), ammonia (□), 1,1,1,2-tetrafluoroethane (Δ), phenol (●).

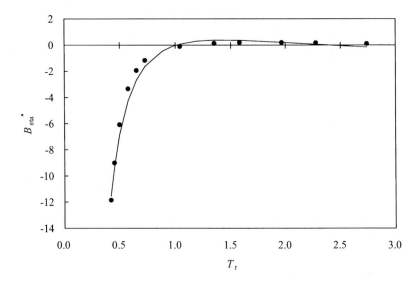

Fig. 7.8 Comparison of the second viscosity virial coefficient data with the values calculated from the extended corresponding-states theory for water.

7.3 Transport Properties in the Critical Region

7.3.1 Critical Nonclassical Behavior of Transport Properties

Sengers (1994) noted that transport properties are related to the dynamic behavior of fluctuations and that it is necessary to consider the structure factor as a function of the time t, $\chi(k,t) = \chi(k)e^{-Dk^2 t}$, where $\chi(k)$ is the static structure factor and D is the diffusivity associated with the order-parameter fluctuations. For a fluid near the vapor-liquid critical point, this diffusivity is to be identified with the thermal diffusivity $D_T = \lambda / \rho c_P$, where λ is the thermal conductivity and c_P the isobaric specific heat capacity. The kinematic viscosity η / ρ, where η is the shear viscosity, is similarly related to the decay rate of the transverse-momentum fluctuations, i.e., fluctuations of the momentum perpendicular to the wave vector. The critical behavior of the diffusivity D_T and, hence, of the thermal conductivity λ, is directly related to the decay of the order-parameter fluctuations. The critical behavior of the viscosity η is caused by a coupling between the order-parameter fluctuations and the viscous momentum fluctuations.

The theory of transport properties of dense fluids is based on the theory of Enskog and its generalizations. The thermal conductivity λ and viscosity η of dense fluids increase generally monotonically with density. In a practical analysis of experimental transport-property data of dense fluids it is often beneficial to consider the excess transport properties:

$$\Delta\lambda(\rho,T) = \lambda(\rho,T) - \lambda_0(T),\tag{7.16}$$

and

$$\Delta\eta(\rho,T) = \eta(\rho,T) - \eta_0(T).\tag{7.17}$$

In Eqs. (7.16) and (7.17), $\lambda_0(T)$ and $\eta_0(T)$ are the thermal conductivity and viscosity in the dilute-gas limit $\rho \to 0$. For simple dense fluids, these residual transport properties are only weakly dependent on the temperature in a substantial range of temperatures. In the classical theory, transport properties are related to the microscopic molecular collisions which are assumed to be effective at short ranges only. As a consequence, the thermal conductivity and viscosity are not affected by the long-range critical correlations. The dependence of the thermal conductivity and viscosity on temperature and density in the critical region is assumed to be qualitatively similar to that observed outside the critical region, and no divergent critical enhancement of these transport properties is expected (Felderhof, 1966). Note that

$$\rho c_{\mathrm{P}} = \rho c_{\mathrm{V}} + T\rho^{-2}(\partial p/\partial T)_p^2 \chi, \qquad (7.18)$$

and the isochoric heat capacity c_{p} diverges at the critical point as the susceptibility χ. Since λ is assumed to remain finite, the thermal diffusivity $D_T = \lambda/\rho c_{\mathrm{p}}$ and, hence, the decay rate Dk^2 for $\chi(k,t)$ vanishes at the critical point as the inverse compressibility χ^{-1}. Thus in the classical theory, the critical slowing down of the fluctuations is a simple consequence of the divergence of the thermodynamic compressibility, while the thermal conductivity and viscosity themselves remain unaffected.

Two theoretical approaches were developed to deal with the critical behavior of dynamic properties. The first is the mode-coupling theory of critical dynamics; the second is the dynamic renormalization-group theory. The predictions of these theories are quite similar, but the mode-coupling theory has been more useful in practice.

To deal with the effects of critical fluctuations, the thermal conductivity λ and the viscosity η are decomposed as:

$$\lambda = \bar{\lambda} + \Delta_c\lambda, \quad \eta = \bar{\eta} + \Delta_c\eta \qquad (7.19)$$

where $\bar{\lambda}$ and $\bar{\eta}$ are background contributions without any effects from critical fluctuations and where $\Delta_c\lambda$ and $\Delta_c\eta$ are enhancements due to the critical fluctuations. The decomposition of λ into $\bar{\lambda}$ and $\Delta_c\lambda$ implies a corresponding decomposition of the thermal diffusivity:

$$D_T = \bar{D}_T + \Delta_c D_T, \qquad (7.20)$$

with $\bar{D}_T = \bar{\lambda}/\rho c_{\mathrm{P}}$. Near the critical point the transport properties not only depend on temperature and density, but also on wave number and frequency. Here we are interested in the transport properties in the long wavelength ($k \to 0$) and zero-frequency limit. In this limit, $\Delta_c D_T$ varies asymptotically as:

$$\Delta_c D_T = \frac{R_{\mathrm{D}} k_{\mathrm{B}} T}{6\pi\eta\xi}, \qquad (7.21)$$

where $R_D = 1.03 \pm 0.03$ is a universal amplitude. As a consequence $\Delta_c \lambda$ becomes:

$$\Delta_c \lambda = \frac{R_D k_B T}{6\pi \eta \xi} \rho c_P . \tag{7.22}$$

Equation (7.21) may be interpreted as the diffusivity of a droplet-like cluster with a radius ξ and with a Stokes law friction coefficient $6\pi \eta \xi$.

The viscosity diverges asymptotically as:

$$\eta = \overline{\eta}(Q\xi)^z , \tag{7.23}$$

where Q is a system-dependent amplitude and z a universal dynamic exponent. The most recent theoretical estimate is:

$$z = 0.063 , \tag{7.24}$$

which is in good agreement with the best experimental results. It follows from Eqs. (7.22) and (7.23) that along the critical isochore above the critical temperature, $\Delta_c \lambda$ diverges as:

$$\lambda = \Lambda_0^+ \left| \Delta \widetilde{T} \right|^{-\psi} , \tag{7.25}$$

with $\psi = \gamma - \nu(1 + z) = 0.57$, while η will diverge as:

$$\eta = H_0^+ \left| \Delta \widetilde{T} \right|^{-\nu z} , \tag{7.26}$$

with $\nu z = 0.040$.

The viscosity η exhibits only a weak-critical divergence. The dynamic renormalization-group theory predicts that η should diverge as ξ^z. The two predictions can be reconciled, if one considers Eq. (7.29) as the first approximation of the power law given by Eq. (7.23), with $z = 8/15\pi^2 = 0.054$ as a first-order estimate for the exponent z, which increases to 0.063.

The range of validity of the asymptotic Eqs. (7.22) and (7.23) for the transport properties is extremely limited. As a consequence, there are hardly any thermal conductivity and viscosity data sufficiently close to the critical point to be represented by these asymptotic equations. The range of densities and temperatures where a critical enhancement of the thermal conductivity occurs quite wide. The critical enhancement of the viscosity is weak and only noticeable in a small range of densities and temperatures near the critical point. Hence, to deal with the actual transport properties of critical fluids, it is necessary to incorporate the nonasymptotic effects from the critical fluctuations on the transport properties.

To obtain a global solution of the mode-coupling equations, it is necessary to retain the finite system-dependent cutoff wave number q_D. Since η and c_V exhibit weak critical anomalies, any wave-number dependence of η and c_V can be neglected. However, the thermal diffusivity D_T can no longer be neglected. One thus obtains:

$$\Delta_c D_T = \frac{\Delta_c \lambda}{\rho c_p} = \frac{R_D kT}{6\pi\eta\xi}(\Omega - \Omega_0) \qquad (7.27)$$

$$\eta = \overline{\eta}\exp(zH), \qquad (7.28)$$

where Ω, Ω_0 and H are functions of temperature and density.

The functions Ω, Ω_0, and H depend on ρ and T, on the specific heat capacities c_p and c_v, on the background transport coefficients $\overline{\lambda}$ and $\overline{\eta}$, on the correlation length ξ and the cutoff parameter q_D. In the limit $q_D\xi \to \infty$ the function Ω - Ω_0 approaches unity and the function H approaches $\ln(Q\xi)$, so that Eqs. (7.21) and (7.23) are recovered for $\Delta_c D_T$ and η. On the other hand, Ω - Ω_0 and H vanish within the limit $q_D\xi \to 0$.

In order to evaluate Eqs. (7.27) and (7.28), the correlation length ξ may be estimated as follows:

$$\xi = \xi_0 \left(\frac{\Delta\widetilde{\chi}}{\Gamma_0^+}\right)^{\frac{v}{\gamma}} \qquad (7.29)$$

with

$$\Delta\widetilde{\chi} = \widetilde{\chi}(\rho,T) - \widetilde{\chi}(\rho,T_r)T_r / T. \qquad (7.30)$$

The amplitude ξ_0 in Eq. (7.29) can be obtained from the equation of state for the free-energy density through Eq. (6.80).

The above equations have been used to represent the transport properties of several fluids in the critical region by Tiesinga et al. (1994) for argon, by Sakonidou et al. (1996) for methane, by Luettmer-Strathmann et al. (1995) for carbon dioxide and ethane, by Krauss et al. (1996) for 1,1-difluoroethane, by Krauss et al. (1993) for 1, 1, 1, 1-tetrafluoroethane, and by Wyczalkowska et al. (2000) for water.

7.3.2 Corresponding States for Transport Properties in the Critical Region

Equations for transport properties in the critical region λ_c and η_c as,

$$y_D = \mathrm{arctg}(q_D \xi) \tag{7.31}$$

$$y_\delta = \frac{\mathrm{arctg}[q_D \xi /(1 + q_D^2 \xi^2)^{1/2}] - y_D}{(1 + q_D^2 \xi^2)^{1/2}} \tag{7.32}$$

$$y_\alpha = \rho k_B T / 8\pi \overline{\eta}^2 \xi \tag{7.33}$$

$$y_\beta = \overline{\lambda}/\overline{\eta}(c_p - c_v) \tag{7.34}$$

$$y_\gamma = c_v /(c_p - c_v) \tag{7.35}$$

$$y_v = y_\gamma y_\delta / y_D \tag{7.36}$$

$$F(x, y_D) = \frac{1}{(1 - x^2)^{1/2}} \ln\left[\frac{1 + x + (1 - x^2)^{1/2} \mathrm{tg}(y_D / 2)}{1 + x - (1 - x^2)^{1/2} \mathrm{tg}(y_D / 2)}\right] \tag{7.37}$$

Equation for thermal conductivity,

$$\lambda_c = \frac{RkT}{6\pi\eta\xi} \rho c_p [\Omega(\{y_i\}) - \Omega_0] \tag{7.38}$$

$$\Omega(\{y_i\}) = \frac{2}{\pi} \frac{1}{1 + y_\gamma} \left[y_D - \sum_1^4 \frac{(a_3 z_i^3 + a_2 z_i^2) + a_1 z_i + a_0}{\Pi_{j=1 j\neq i}^4 (z_i - z_j)} F(z_i, y_D)\right] \tag{7.39}$$

$$\Omega_0 = \frac{1 - \exp\left\{-[(q_D \xi)^{-1} + (q_D \xi \rho_c / \rho)^2 / 3]^{-1}\right\}}{\pi / 2[1 + y_a(y_D + y_\delta) + y_\beta(1 + y_\gamma)^{-1}]} \tag{7.40}$$

$$\prod_{i=1}^4 (z + z_i) = z^4 + b_3 z^3 + b_2 z^2 + b_1 z + b_0 = 0 \tag{7.41}$$

$$a_0 = y_\gamma^2 - y_\alpha y_\gamma y_\delta \qquad (7.42)$$

$$b_0 = y_\alpha y_\gamma y_\delta \qquad (7.43)$$

$$a_1 = y_\alpha y_\gamma y_D \qquad (7.44)$$

$$b_1 = y_\alpha y_\gamma y_D \qquad (7.45)$$

$$a_2 = y_\gamma - y_\beta - y_\alpha y_\delta \qquad (7.46)$$

$$b_2 = y_\gamma + y_\beta + y_\alpha y_\delta \qquad (7.47)$$

$$a_3 = y_\alpha y_D \qquad (7.48)$$

$$b_3 = y_\alpha y_D \qquad (7.49)$$

Equation for viscosity,

$$\eta_c = \bar{\eta}\left[e^{zH(\{y_1\})} - 1 \right] \qquad (7.50)$$

$$H(\{y_i\}) = h(\{y_i\}) + \sum_{i=1}^{3} \frac{\left(c_2 v_i^2 + c_1 v_i + c_0 \right)}{\prod_{j=1 j \neq i}^{3} \left(v_i - v_j \right)} F(v_i, y_D) \qquad (7.51)$$

$$h(\{y_i\}) = \left[3 y_\gamma y_\eta + 3 y_\eta / 2 - y_\eta^3 - y_\nu \right] y_D + \left[y_\eta^2 - 2 y_\gamma - 5/4 \right] \sin y_D$$
$$- y_\eta (\sin 2 y_D)/4 + (\sin 3 y_D)/12 + \frac{\left[y_\gamma (1 + y_\gamma) \right]^{3/2}}{\left(y_\nu - y_\gamma y_\eta \right)} \arctan\left(\left[y_\gamma (1 + y_\gamma) \right]^{1/2} \tan y_D \right)$$
$$\qquad (7.52)$$

$$\prod_{i=1}^{3} (v + v_i) = v^3 + y_\eta v^2 + y_\gamma v + y_\nu = 0 \qquad (7.53)$$

$$c_0 = y_\eta y_\nu \left(y_\eta^2 - 3 y_\gamma - 2 \right) + y_\nu^2 - y_\gamma y_\nu \left(1 + y_\gamma \right)^2 / \left(y_\nu - y_\gamma y_\eta \right) \qquad (7.54)$$

$$c_1 = \left(y_\nu - y_\gamma y_\eta \right)\left(y_\eta^2 - 3 y_\gamma - 2 \right) + y_\gamma^2 \left(1 + y_\gamma \right)^2 / \left(y_\nu - y_\gamma y_\eta \right) \qquad (7.55)$$

$$c_2 = y_\eta^4 - 4 y_\gamma y_\eta^2 - 2 y_\eta^2 + 2 y_\eta y_\nu + 3 y_\gamma^2 + 4 y_\gamma + 1 \qquad (7.56)$$

Equation for the correlation length:

$$\xi = \xi_0 \Gamma^{-\nu/\eta} \left[\tilde{X}(T,\rho) - \tilde{X}(T_r,\rho)\frac{T_r}{T} \right]^{\nu/\gamma} \tag{7.57}$$

where $\gamma = \nu(2-\eta)$, $T_r = 2T_c$, $z=0.063$, $\nu = 0.630$, $\gamma = 1.239$

Critical amplitudes: $R=1.03$. These parameters are correlated from the extended corresponding-states theory as,

$$\xi_0 /(V_c / N_A)^{1/3} = 0.327 - 0.219\omega + 11.19\theta \tag{7.58}$$

$$\Gamma = 0.06103 - 0.05805\omega + 4.441\theta \tag{7.59}$$

Cutoff wave number:

$$q_D^{-1} /(V_c / N_A)^{1/3} = 0.3804701 - 0.1483888\omega + 11.965591\theta \tag{7.60}$$

7.4 Viscosity in the Entire Fluid Region

7.4.1 Equation for Viscosity

For the high-density dense state, the most successful theoretical model to predict the effect of density on the viscosity is due to Enskog, and the resulting equation for the viscosity of the hard sphere is (Chapman and Cowling, 1952, 1970; Hirschfelder et al., 1954)

$$\eta_E / \eta_0 = 1/ g(\sigma) + 0.8b\rho + 0.761g(\sigma)(b\rho)^2 , \tag{7.61}$$

where $g(\sigma)$ is the radial distribution function at contact (Carnaham and Starling, 1969; Dymond, 1985), and

$$g(\sigma) = (1 - 0.5\xi)/(1-\xi)^3 , \tag{7.62}$$

where $\eta_0 = 5(mkT)^{1/2}/16\pi^{1/2}\sigma^2$, $b = 2\pi N_A \sigma^3 /3$, $\xi = b\rho/4$. ρ is the molar density. Eq. (7.64) requires the known parameter σ along with the density and molar mass. The parameter σ of the hard sphere used in Eq. (7.62) is the same as that of the Lennard-Jones potential (Dymond, 1985; Heyes, 1988).

For a real fluid, η_0 in Eq. (7.64) can be corrected by η_{CE} in Eq. (7.1), in that the Lennard-Jones intermolecular potential is more reasonable than the hard sphere in the low-density gas phase because in the Lennard-Jones potential the derived properties are dependent mainly on the attractive part of the potential. Consequently, the reduced viscosity was defined as

$$H = \eta_{CE}(\eta_E / \eta_0), \tag{7.63}$$

where H is a quantity to reduce the viscosity. The following simple equation for the viscosity of a fluid was proposed as:

$$\eta_r = \eta / H = X + (1 - X)(c_0 + c_1 / T_r^3 + c_1 / T_r^8), \tag{7.64}$$

where η is the viscosity and η_r reduced viscosity. η_r may be almost density-independent; however it is temperature-dependent in the low-temperature liquid region. The function $X = \exp\left[-(2\rho_r)^3(2\rho_r - 1)^3\right]$ is a crossover function from high density to low density (Xiang, 1995). c_0, c_1, and c_2 are the three substance-dependent parameters.

7.4.2 Corresponding-States Viscosity

According to the extended corresponding-states theory, the three substance-dependent parameters in Eq. (7.64) may be obtained as follows:

$$c_0 = c_{00} + c_{01}\omega + c_{02}\theta,$$
$$c_1 = c_{10} + c_{11}\omega + c_{12}\theta,$$
$$c_2 = c_{20} + c_{21}\omega + c_{22}\theta \tag{7.65}$$

The coefficients c_{ij} of Eq. (7.65) are determined by fitting the saturated-liquid viscosity data for argon, methane, ethane, carbon dioxide, propane, 1,1,1,2-tetrafluoroethane (R134a), 1,1-difluoroethane (R152a), ammonia, and water, which represent typical simple, nonpolar and polar substances. A wide range of experimental data is available for these substances, both in the liquid state and in the zero-density limit, along with their critical parameters, the acentric factors, Lennard-Jones parameters, and data sources for zero-density viscosity and saturated liquid viscosity, as listed in Tables 7.2, 7.6, and 7.7. It should be noted that these data calculated from the recommended correlation equations usually contain a few percentages for the accuracy of used data. The parameters c_{ij} given in Table 7.8 are independent of the temperature, the density, and the specific substance.

7.4.3 Comparison with Experimental Data

The determination of the Lennard-Jones parameters for the substances listed in Table 7.7 is based on the reasonable agreement of calculated results with the experimental viscosity data both in the zero-density limit and in the saturated liquid state. The present deviations with the determined Lennard-Jones parameters from the experimental viscosity data in the zero-density limit are usually within 1% to 2% as shown in Fig. 7.9, which shows that the parameters are a good representation of the dilute viscosity data, and fit reasonably well within the experimental uncertainties for non-polar and weakly polar substances. For strongly polar substances the results are also in a good agreement in the usual temperature range at reduced temperatures of about 0.7. Although the deviations of methane at low temperatures and of argon at high temperatures are relatively large, within 3 % to 4%, the Lennard-Jones parameters are in good agreement with those in the literature, as listed in Table 7.7. Generally, the parameters agree with the dilute viscosity data sets with an error of 1% to 3%. The fact that the available data and the determined Lennard-Jones parameters are in good agreement proves that reasonable Lennard-Jones parameters can be determined from a few experimental viscosity data points both in the dilute and in the liquid phase to fit the extended corresponding-state method. The viscosity over the entire fluid surface can be predicted from the determined Lennard-Jones parameters along with other corresponding-states parameters. It should be noted that the Lennard-Jones parameter σ can affect mainly the calculation result of liquid viscosity and less that of the dilute gas viscosity. Because the Lennard-Jones parameters are only substance-dependent, in principle, they can also be determined from other methods; however, the Lennard-Jones parameters are determined to use experimental gas and liquid viscosity data to improve the predictive accuracy if any reliable data are available. In fact, as shown in Table 7.7 some existing Lennard-Jones parameters not determined from the viscosity are also in good agreement with the parameters determined here.

For the saturated liquid state, Figs. 7.10 to 7.12 show how the extended corresponding-states theory represents recommended saturated-liquid viscosity data for various fluids from the triple point to the critical point along the vapor-liquid coexistence curve. For most cases the agreement between the present method and the data is generally comparable to the agreement between the different sets of data with a 5% error far from the critical point and sometimes 10% near the critical point. In the liquid phases this property is also subject to large uncertainties; in general, the viscosity has a few percentages in experimental uncertainty. The present method gives a very good description within a wide range of temperature and density and also can describe the viscosities of polar, hydrobonding, and associating substances, such as chlorodifluoromethane (R22), difluoromethane (R32), 1, 1-dichloro-2, 2, 2-trifluoroethane (R123), 1, 1, 1, 2-tetrafluoroethane (R134a), 1, 1-difluoroethane (R152a), water, methanol, ammonia, and acetone.

Figs. 7.10 to 7.12 show a predictive comparison of nitrogen, ethylene, benzene, octane, R22, R32, R123, methanol, and acetone in the saturated liquid state. Figs. 7.10 to 7.12 show that the present method represents recommended viscosity data for ethane (Friend et al., 1991; Hendl et al., 1994) and water (Sengers and Watson, 1986) over their entire fluid states from the dilute state to a highly compressed liquid state. Since the reduced viscosity state is nearly density-independent, the prediction of density-dependent viscosity in the low-temperature liquid state and in the supercritical region can be well obtained. The available experimental data for the comparisons in Figs. 7.13 to 7.16 cover their entire fluid surfaces at temperatures from the triple point to high temperatures (1073 K for water and 500 K for ethane) and at pressures up to 100 MPa for water and 70 MPa for ethane, and at densities from the ideal gas to the liquid region. It can be seen that for the entire fluid surface the predictive result is usually good, with a deviation of less than 10% at most, and usually within 5 % or less. At the triple-point isotherm, the reduced viscosity is not very density-independent, which causes a relatively large deviation. This probably indicates that as the fluid state approaches the solid phase, the hard-sphere model is less applicable than for the normal liquid state. The extended corresponding-states method with reduced viscosity compensates for the absence of a rigorous theory for the viscosity of a fluid in the intermediate-density range, with a deviation of 6 to 12 %. In the critical region, there is usually 10% experimental accuracy. It should be noted that the coefficients of Eq. (7.65) are determined only from the saturated-liquid viscosity data and that water is a strongly hydrobonding associating polar fluid and ethane a simple nonpolar hydrocarbon substance.

The simple and accurate corresponding-states method was developed to correlate and predict the viscosity of a real fluid including highly polar substances within the corresponding-states framework. Based on the Chapman-Enskog theory in the dilute vapor and in the dense hard-sphere liquid state, the reduced viscosity reflects the physical behavior of the viscosity of a real fluid in the dilute vapor and the high-density liquid state. The prediction of the viscosity over the entire region from triple point to high temperature and ideal gas to high densities from the present method is more reliable, accurate, and consistent than existing methods. Comparison with the experimental viscosity data shows that the present method provides a reliable means to account for the non-hard-spherical effect from the hard-sphere model on the viscosity of a real fluid. For viscosity, which has not been extensively studied and is a property that is not well understood, the present method makes it possible to improve the theories, to understand the intermolecular potential and to test the reliability of experimental data.

Table 7.6 Data Sources for Zero-Density and Saturated Liquid Viscosities

Substance	Refs.(0)	Refs.(SL)
Argon	Kestin et al., 1984	Yonglove and Hanley, 1986
Nitrogen	Boushehri et al., 1987; Millat and Wakeham, 1989	Stephan et al., 1987
Methane	Boushehri et al., 1987; Trengore and Wakeham, 1987	Haynes, 1973
Carbon dioxide	Boushehri et al., 1987; Trengore and Wakeham, 1987	Fenghour et al., 1998
Ethylene	Boushehri et al., 1987	Holland et al., 1983
Ethane	Hendl et al., 1991	Friend et al., 1991
Propane	Vogel, 1995	Holland et al., 1979; Diller, 1982
Benzene	Vogel et al., 1986	Dymond and Robertson, 1985; Dymond and Young, 1981; Knapstad et al., 1989
Octane		Knapstad et al., 1989; Dymond and Young, 1980
Chlorodifluoromethane	Kestin and Wakeham, 1979; Takahashi et al., 1983	Assael and Polimatidou, 1994; Diller et al., 1992
Difluoromethane	Takahashi et al., 1995	Ripple and Matar, 1993; Assael et al., 1994; Oliveira and Wakeham, 1993
2,2-Dichloro-1,1,1-trifluoroethane	Nabizadeh and Mayinger, 1992	Kumagai and Takahashi, 1991; Diller et al., 1993
1,1,1,2-Tetrafluoroethane	Nabizadeh and Mayinger, 1992; Wilhelm and Vogel, 1996	Ripple and Matar, 1993; Assael et al., 1994; Diller et al.,1993; Oliveira and Wakeham, 1993
1,1-Difluoroethane	Krauss et al., 1996	Assael et al., 1994; Van der Gulik, 1995
Ammonia	Fenghour et al., 1995	Fenghour et al.,1995
Water	Sengers and Watson, 1986	Sengers and Watson, 1986
Methanol	Vogel et al., 1986	Isdale et al., 1985; Tanaka et al., 1987; Assael and Polimatidou, 1994
Acetone		Ovchinnikova et al., 1983

Table 7.7 Lennard-Jones Potential Parameters σ and ε/k

Substance	σ (nm)/ ε/k (K)	σ_{LJ} (nm)	$(\varepsilon/k)_{LJ}$ (K)
Argon	0.3465/113.5(Mourits and Rummens, 1977); 0.341/117(Ben-Amotz and Herschbach, 1990)	0.347	120
Nitrogen	0.3738/82.0(Mourits and Rummens, 1977); 0.3652/98.4(Boushehri et al., 1987)	0.365	98
Methane	0.379/142.1(Mourits and Rummens, 1977); 0.373/142(Ben-Amotz and Herschbach, 1990); 0.3721/161.4(Boushehri et al., 1987); 0.37238/150(Sun and Teja, 1998)	0.373	160
Carbon dioxide	0.3943/200.9(Mourits and Rummens, 1977); 0.369/247(Ben-Amotz and Herschbach, 1990); 0.3769/245.3(Boushehri et al., 1987)	0.373	260
Ethylene	0.4155/225.6(Mourits and Rummens, 1977); 0.413/230(Ben-Amotz and Herschbach, 1990); 0.407/244.3(Boushehri et al., 1987)	0.415	225
Ethane	0.4407/227.9(Mourits and Rummens, 1977); 0.429/244(Ben-Amotz and Herschbach, 1990); 0.437/241.9(Boushehri et al., 1987); 0.4243/249(Sun and Teja, 1998); 0.43075/264.7(Friend et al., 1991; Hendl et al., 1994)	0.430	260
Propane	0.5114/237.2(Mourits and Rummens, 1977); 0.476/316(Ben-Amotz and Herschbach, 1990); 0.47069/306(Sun and Teja, 1998); 0.4915/285.6(Vogel, 1995)	0.480	320
Benzene	0.5455/401.2(Mourits and Rummens, 1977); 0.515/557(Ben-Amotz and Herschbach, 1990)	0.528	440
Octane	0.7024/357.7(Mourits and Rummens, 1977); 0.635/617(Ben-Amotz and Herschbach, 1990)	0.648	600
Chlorodifluoromethane		0.450	320
Difluoromethane		0.390	350
2,2-Dichloro-1,1,1-trifluoroethane	0.5909/275.16(Nabizadeh and Mayinger, 1992)	0.550	360
1,1,1,2-Tetra	0.5067/277.74(Nabizadeh and Mayinger,	0.490	325

fluoroethane	1992); 0.506/288.8(Wilhelm and Vogel, 1996)		
1,1-Difluoro ethane	0.46115/354.84(Krauss et al., 1996)	0.460	355
Ammonia	0.3215/309.9(Mourits and Rummens, 1977)	0.325	360
Water	0.271/506(Mourits and Rummens, 1977); 0.289/371(Ben-Amotz and Herschbach, 1990); 0.3034/630(Sun and Teja, 1998)	0.280	680
Methanol	0.3657/385.2(Mourits and Rummens, 1977); 0.380/376(Ben-Amotz and Herschbach, 1990); 0.3933/520(Sun and Teja, 1998)	0.397	340
Acetone	0.4599/458(Mourits and Rummens, 1977); 0.476/467(Ben-Amotz and Herschbach, 1990)	0.476	467

Table 7.8 General Coefficients of Eq. (7.65)

c_{00}	0.82987	c_{10}	0.345036	c_{20}	-0.001241592
c_{01}	0.583014	c_{11}	-0.242490	c_{21}	0.007206665
c_{02}	9.15963	c_{12}	-36.3136	c_{22}	4.39426

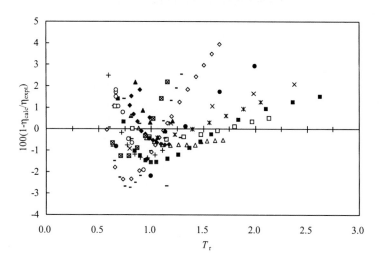

Fig. 7.9 Comparison of the viscosity data in the zero-density limit for substances with values calculated from Eq. (7.1). ($\bullet$) argon; ($\times$) nitrogen; (+) methane; ($\square$) ethane; (Δ) propane; (*) carbon dioxide; ($\blacksquare$) benzene; ($-$) R22; ($\blacktriangle$) R32; ($\circ$) R123; ($\blacklozenge$) R134a; (-) ammonia; ($\Diamond$) water; and ($\square\times$) methanol. [Reproduced with permission from Xiang (2003)]

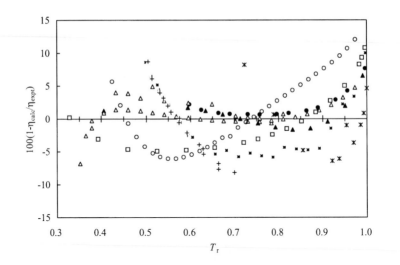

Fig. 7.10 Comparison of the saturated-liquid viscosity data for nonpolar substances with values calculated from the extended corresponding-states theory, Eq. (7.64). (×) argon; (▲) nitrogen; (■) methane; (○) ethylene; (□) ethane; (Δ) propane; (*) Carbon dioxide; and (+) Benzene. [Reproduced with permission from Xiang (2003)]

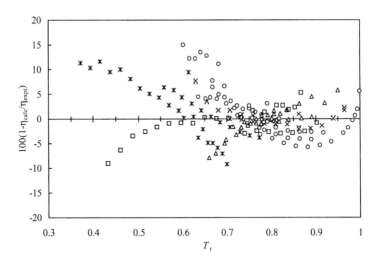

Fig. 7.11 Comparison of the saturated-liquid viscosity data for some highly polar substances with values calculated from the extended corresponding-states theory, Eq. (7.64).   (□)   chlorodifluoromethane;   (Δ)   difluoromethane;   (*) 1,1-dichloro-2,2,2-trifluoroethane;   (○)   1,1,1,2-tetrafluoroethane;   and   (×) 1,1-difluoroethane(R152a). [Reproduced with permission from Xiang (2003)]

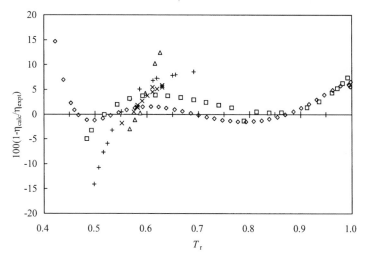

Fig. 7.12 Comparison of the saturated-liquid viscosity data for some highly polar substances with values calculated from the extended corresponding-states theory, Eq. (7.64). (Δ) octane; (◊) water; (□) ammonia;   (×) methanol; and (+) acetone. [Reproduced with permission from Xiang (2003)]

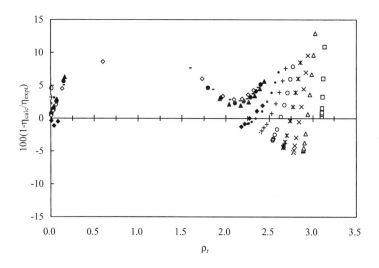

Fig. 7.13 Comparison of the viscosity data over the entire fluid state for ethane with values calculated from the extended corresponding-states theory, Eq. (7.64), below the critical isotherm. (□) (T_r 0.328); (Δ) 0.459; (×) 0.524; (*) 0.590; (○) 0.655; (+) 0.721; (■) 0.786; (♦) 0.819; (▲) 0.852; (●) 0.917; (-) 0.950; and (◊) 0.983. [Reproduced with permission from Xiang (2003)]

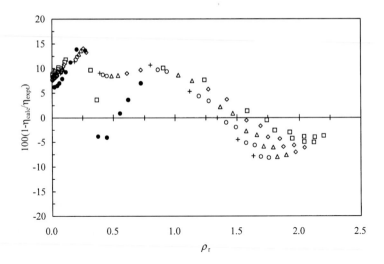

Fig. 7.14 Comparison of the viscosity data over the entire fluid state for ethane with values calculated from the extended corresponding-states theory, Eq. (7.64), above the critical isotherm. (•) T_r 1.015; (♦) 1.048; (□) 1.114; (×) 1.179; (◊) 1.245-1.310; (Δ) 1.376-1.441; (○) 1.507-1.572; and (+) 1.638. [Reproduced with permission from Xiang (2003)]

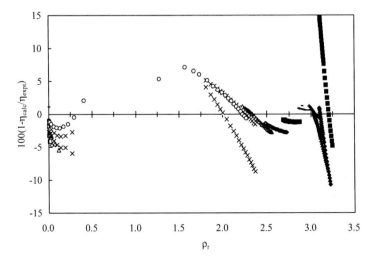

Fig. 7.15 Comparison of the viscosity data over the entire fluid state for water with values calculated from the extended corresponding-states theory, Eq. (7.64), below the critical isotherm. (+) T_r 0.422; (♦) 0.461; (▲) 0.499; (*) 0.538; (—) 0.577; (-) 0.654; (□) (0.731); (◊) 0.808; (Δ) 0.886; (×) 0.963; and (○) 1.002. [Reproduced with permission from Xiang (2003)]

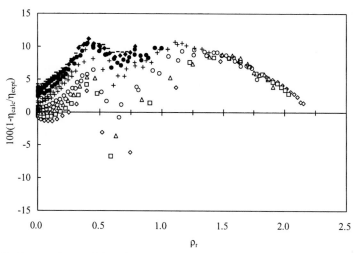

Fig. 7.16 Comparison of the viscosity data over the entire fluid state for water with values calculated from the extended corresponding-states theory, Eq. (7.64), above the critical isotherm. (◊) T_r, 1.04; (□) 1.079 (Δ) 1.118; (○) 1.156 and 1.195; (+) 1.272; (--) 1.349; (●) 1.42 and 1.504 (♦) 1.581; and (×) 1.658. [Reproduced with permission from Xiang (2003)]

References

Alexandrov, A. A., A. A. Ivanov, and A. B. Matveev, 1975, *Teploenergetika* **22(4)**: 59.

Assael, M. J., H. J. Dymond, and S. K. Polimatidou, 1994, *Int. J. Thermophys.* **15**: 591.

Assael, M. J. and S. K. Polimatidou, 1994, *Int. J. Thermophys.* **15**: 95.

Assael, M. J. and S. K. Polimatidou, 1994, *Int. J. Thermophys.* **15**: 779.

Assael, M. J., S. K. Polimatidou, E. Vogel, and W. A. Wakeham, 1994, *Int. J. Thermophys.* **15**: 575.

Aziz, R. A. and M. J. Slaman, 1986, *Mol. Phys.* **58**: 679.

Ben-Amotz, D. and D. R. Herschbach, 1990, *J. Phys. Chem.* **94**: 1038.

Bich, E., J. Millat, and E. Vogel, 1990, *J. Phys. Chem. Ref. Data* **19**: 1289.

Bich, E., J. Millat, and E. Vogel, 1987, *Wiss. Z. W.-Pieck-Univ. Rostock* **36(8)**: 5.

Boushehri, A., J. Bzowski, J. Kestin, and E. A. Mason, 1987, *J. Phys. Chem. Ref. Data* **16**: 445.

Brule, M. R. and K. E. Starling, 1984, *Ind. Eng. Chem. Process Des.Dev.* **23**: 833.

Carnaham, N. F. and K. E. Starling, 1969, *J. Chem. Phys.* **51**: 635.

Chapman, S. and T. G. Cowling, 1939, The Mathematical Theory of Nonuniform Gases (Cambridge University Press, New York).

Chapman, S. and T. G. Cowling, 1952, The Mathematical Theory of Nonuniform

Gases, 2nd ed. (Cambridge University Press, New York).

Chapman, S. and T. G. Cowling, 1970, The Mathematical Theory of Nonuniform Gases, 3rd ed. (Cambridge University Press, New York).

Chung, T. H., L. L. Lee, and K. E. Starling, 1984, *Ind. Eng. Chem. Fundam.* **23**: 8.

Diller, D. E., 1982, *J. Chem. Eng. Data* **27**: 240.

Diller, D. E., A. S. Aragon, and A. Laesecke, 1992, *Int. J. Refrig.* **16**: 19.

Diller, D. E., A. S. Aragon, and A. Laesecke, 1993, *Fluid Phase Equil.* **88**: 251.

Dowdell, D. C. and G. P. Matthews, 1993, *J. Chem. Soc. Faraday Trans.* **89**: 3545.

Dymond, J. H., 1985, *Rev. Chem. Soc.* **14**: 317.

Dymond, J. H., E. Bich, E. Vogel, W. A. Wakeham, V. Vesovic, and M. J. Assael, 1996, in Transport Properties of Fluids: Their Correlation, Prediction and Estimation, J. Millat, J. H. Dymond, and C. A. Nieto de Castro, eds. (Cambridge University Press, New York).

Dymond, J. H. and J. Robertson, 1985, *Int. J. Thermophys.* **6**: 21.

Dymond, J. H. and K. J. Young, 1980, *Int. J. Thermophys.* **1**: 331.

Dymond, J. H. and K. J. Young, 1981, *Int. J. Thermophys.* **2**: 237.

Fenghour, A., W. A. Wakeham, and V. Vesovic, 1998, *J. Phys. Chem. Ref. Data* **27**: 31.

Fenghour, A., W. A. Wakeham, V. Vesovic, J. T. R. Watson, J. Millat, and E. Vogel, 1995, *J. Phys. Chem. Ref. Data* **24**: 1649.

Friend, D. G., H. Ingham, and J. F. Ely, 1991, *J. Phys. Chem. Ref. Data* **20**: 275.

Friend, D. G. and J. C. Rainwater, 1984, *Chem. Phys. Lett.* **107**: 590.

Gracki, J. A., G. P. Flynn, and J. Ross, 1969, *J. Chem. Phys.* **51**: 3856.

Gross, U. and Y. W. Song, 1996, *Int. J. Thermophys.* **17**: 607.

Haynes, W. M., 1973, *Physica* **70**: 410.

Hendl, S., J. Millat, V. Vesovic, E. Vogel, and W. A. Wakeham, 1991, *Int. J. Thermophys.* **12**: 999.

Hendl, S., J. Millat, E. Vogel, V. Vesovic, W. A. Wakeham, J. Luettmer-Strathmann, J. V. Sengers, and M. J. Assael, 1994, *Int. J. Thermophys.* **15**: 1.

Hendl, S. and E. Vogel, 1992, *Fluid Phase Equil.* **76**: 259.

Heyes, D. M., 1988, *Chem. Phys. Lett.* **153**: 319.

Hirschfelder, J. O., C. F. Curtiss, and R. B. Bird, 1964, The Molecular Theory of Gases and Liquids (John Wiley, New York).

Holland, P. M., B. E. Eaton, and H. J. M. Hanley, 1983, *J. Phys. Chem. Ref. Data* **12**: 917.

Holland, P. M., H. J. M. Hanley, K. E. Gubbins, and J. M. Haile, 1979, *J. Phys. Chem. Ref. Data* **8**: 559.

Huber, M. L. and H. J. Hanley, 1996, in Transport Properties of Fluids: Their Correlation, Prediction and Estimation, J. Millat, J. H. Dymond, and C. A. Nieto de Castro, eds. (Cambridge University Press, New York).

Isdale, J. D., A. J. Easteal, and L. A. Woolf, 1985, *Int. J. Thermophys.* **6**: 439.

Iwasaki, H. and M. Takahashi, 1968, *Rev. Phys. Chem. Jpn.* **38**: 18.

Iwasaki, H. and M. Takahashi, 1975, *Bull. Chem. Soc. Jpn.* **48**: 988.

Kestin, J., K. Knierim, E. A. Mason, B. Najafi, S. T. Ro, and M. Waldman, 1984, *J. Phys. Chem. Ref. Data* **13**: 229.

Kestin, J., O. Korfali, J. V. Sengers, and B. Kamgar-Parsi, 1981, *Physica A* **106**: 415.

Kestin, J., E. Paykoc, and J. V. Sengers, 1971, *Physica* **54**: 1.

Kestin, J. and W. A. Wakeham, 1979, Ber. Bunsenges. *Phys. Chem.* **83**: 573.

Knapstad, B., P. A. Skjolsvik, and H. A. Oye, 1989, *J. Chem. Eng. Data* **34**: 37.

Krauss, R., J. Luettmer-Strathmann, J. V. Sengers, and K. Stephan, 1993, *Int. J. Thermophys.* **14**: 951.

Krauss, R., V. C. Weiss, T. A. Edison, J. V. Sengers, and K. Stephan, 1996, *Int. J. Thermophys.* **17**: 731.

Kuchenmeister, C. and E. Vogel, 2000, *Int. J. Thermophys.* **21**: 329.

Kuchenmeister, C. and E. Vogel, 1998, *Int. J. Thermophys.* **19**: 1085.

Kumagai, A. and S. Takahashi, 1991, *Int. J. Thermophys.* **12**: 105.

Laesecke, A., R. Krauss, K. Stephan, and W. Wagner, 1990, *J. Phys. Chem. Ref. Data* **19**: 1089.

Leland, T. W. and P. S. Chappelear, 1968, *Ind. Eng. Chem.* **60 (7)**: 15.

Luettmer-Strathmann, J., J. V. Sengers, and G. A. Olchowy, 1995, *J. Chem. Phys.* **103**: 7482.

Mason, E. A. and L. Monchick, 1962, *J. Chem. Phys.* **36**: 1622.

Millat, J. and W. A. Wakeham, 1989, *J. Phys. Chem. Ref. Data* **18**: 565.

Mourits, F. M. and F. H. Rummens, 1977, *Can. J. Chem.* **55**: 3007.

Nabizadeh, H. and F. Mayinger, 1992, *High Temp. High Press.* **24**: 221.

Neufeld, P. D., A. R. Janzen, and R. A. Aziz, 1972, *J. Chem. Phys.* **57**: 1100.

Oliveira, C. M. B. P. and W. A. Wakeham, 1993, *Int. J. Thermophys.* **14**: 1131.

Oliveira, C. M. B. P. and W. A. Wakeham, 1993, *Int. J. Thermophys.* **14**: 33.

Ovchinnikova, R. A., N. V. Denkina, and A. V. Porai-Koshits, 1983, *J. Appl. Chem. USSR,* **55**: 1737.

Pitzer, K. S. and R. F. Curl, 1957, *J. Am. Chem. Soc.* **79**: 2369.

Pitzer, K. S., D. Z. Lippmann, R. F. Curl, Jr., C. M. Huggins, and D. E. Petersen, 1955, *J. Am. Chem. Soc.* **77**: 3433.

Poling, B. E., J. M. Prausnitz, and J. P. O'Connell, 2001, The Properties of Gases and Liquids, 5th ed. (McGraw-Hill, New York).

Rainwater, J. C., 1984, *J. Chem. Phys.* **81**: 495.

Rainwater, J. C. and D. G. Friend, 1987, *Phys. Rev. A* **36**: 4062.

Reid, R. C., J. M. Prausnitz, and B. E. Poling, 1987, The Properties of Gases and Liquids, 4th ed. (McGraw-Hill, New York).

Reid, R. C., J. M. Prausnitz, and T. K. Sherwood, 1977, The Properties of Gases and Liquids, 3rd ed. (McGraw-Hill, New York).

Ripple, D. and O. Matar, 1993, *J. Chem. Eng. Data* **38**: 560.

Ross, M., R. Szezepanski, R. D. Trengove, and W. A. Wakeham, 1986, presented at the AIChE Winter Meeting, Miami.

Sakonidou, E. P., H. R. Van den Berg, C. A. ten Seldam, and J. V. Sengers, 1996, *J.*

Chem. Phys. **105**: 10535.

Sengers, J. V., 1994, in Supercritical Fluids, E. Kiran and J. M. H. Levelt Sengers eds. (Kluwer, Dordrecht), pp 231-271.

Sengers, J. V. and J. T. R. Watson, 1986, *J. Phys. Chem. Ref. Data* **15**: 1291.

Stephan, K., R. Krauss, and A. Laesecke, 1987, *J. Phys. Chem. Ref. Data* **15**: 1323.

Strehlow, T. and E. Vogel, 1989, *Physica* **161A**: 101.

Sun, T. F. and A. S. Teja, 1998, *Ind. Eng. Chem. Res.* **37**: 3151.

Takahashi, M., C. Yokoyama, and S. Takahashi, 1987, *J. Chem. Eng. Data* **32**: 98.

Takahashi, M., N. Shibasaki-Kitakawa, Ch. Yokoyama, and S. Takahashi, 1995, *J. Chem. Eng. Data* **40**: 900.

Takahashi, M., S. Takahashi, and H. Iwasaki, 1983, *Kagaku Kogaku Ronbunshu* **9**: 482.

Tanaka, Y., Y. Matsuda, H. Fujiwara, H. Kubota, and T. Matiza, 1987, *Int. J. Thermophys.* **8**: 147.

Taxman, N., 1958, *Phys. Rev.* **110**: 1235.

Tiesinga, B. W., J. Luettmer-Strathmann, and J. V. Sengers, 1994, *J. Chem. Phys.* **101**: 6944.

Trengore, R. D. and W. A. Wakeham, 1987, *J. Phys. Chem. Ref. Data* **16**: 175.

Uribe, F. J., E. A. Mason, and J. Kestin, 1990, *J. Phys. Chem. Ref. Data* **19**: 1123.

Van der Gulik, P. S., 1995, *Int. J. Thermophys.* **16**: 867.

Vargeftik, N. B., 1975, Tables on the Thermophysical Properties of Liquids and Gases (Wiley, New York).

Vargeftik, N. B., 1993, Handbook of Thermal Conductivity of Liquids and Gases, (CRC Press, Boca Raton).

Vogel, E., 1995, *Int. J. Thermophys.* **16**: 1335.

Vogel, E., E. Bich, and R. Nimz, 1986, *Physica* **139A**: 188.

Vogel, E. and S. Hendl, 1992, *Fluid Phase Equil.* **79**: 313.

Vogel, E., B. Holdt, and T. Strehlow, 1988, *Physica* **148A**: 46.

Vogel, E. and A. K. Neumann, 1993, *Int. J. Thermophys.* **14**: 805.

Vogel, E. and T. Strehlow, 1988, *Z. Phys. Chem. Leipzig* **269**: 897.

Wang Chang, C. S. and G. E. Uhlenbeck, 1951, Transport Phenomena in Polyatomic Molecules (Michigan Univ. Eng. Research Inst. Report CM-681).

Wilhelm, J. and E. Vogel, 1996, *Fluid Phase Equil.* **125**: 257.

Wyczalkowska, A. K., Kh. S. Abdulkadirova, M. A. Anisimov, and J. V. Sengers, 2000, *J. Chem. Phys.* **113**: 4985.

Wyczalkowska, A. K. and J. V. Sengers, 1999, *J. Chem. Phys.* **111**: 1551.

Xiang, H. W., 1995, New Vapor-Pressure Equation and Crossover Equation of State, Doctoral Dissertation (Jiaotong University, Xi'an).

Xiang, H. W., 2001, *Fluid Phase Equil.* **187**: 221.

Xiang, H. W., 2003, Thermophysicochemical Properties of Fluids: Corresponding-States Principle and Practise (Science Press, Beijing), in Chinese.

Younglove, B. A. and H. J. M. Hanley, 1986, *J. Phys. Chem. Ref. Data* **15**: 1323.

Chapter 8

Surface Tension

8.1 Introduction

The surface tension is an important property for many processes and physical phenomena. However, experimental data are usually scarce and in limited temperature ranges. The surface tension is also arguably the most important of the inhomogeneous fluid properties. It reflects the range of interactions in a fluid more directly than the bulk properties do. Modern liquid theory has brought much insight into the mechanism of surface tension, and simulations have added further insights and quantitative predictions at least for simple models for fluids such as the Lennard-Jones fluids. However, the sensitivity to long-range interactions coupled with the low symmetry due to the interface complicates the application of both theory and simulation to surface tension, particularly in complex fluids.

The thermodynamics of surfaces forms a fascinating field for study. Many mathematical physical methods have been proposed that differ considerably but can be reduced to similar equations relating macroscopically measurable quantities. Two different ways to determine the surface tension of a liquid were developed in the nineteenth century. Rowlinson reviewed the progress in this field (Van der Waals, 1873). Young and Laplace obtained an expression for the surface tension of a liquid bounded by a sharp surface at which the liquid density falls abruptly to that of the gas. Later in the century, Fuchs, Rayleigh and most convincingly, Van der Waals, developed the theory for a diffuse interface in which there was a smooth gradient of density at each height.

Van der Waals (1894) suggested a theory of surface tension that showed the relation of thermodynamic properties to the coexisting phase surface of fluids and found that the surface tension $\sigma / p_c^{2/3} T_c^{1/3}$ can be correlated with $1 - T_r$ in the critical region. Macleod (1923) suggested a simple relation that showed that the surface tension of liquids is a function of the difference between the densities of coexisting liquid and vapor phases, which may be used for many substances in a wide range of temperature. Since the corresponding-states principle was derived from the equation of state, the surface tension is expected to follow the corresponding-states principle as well, according to the Van der Waals theory.

The corresponding-states principle has been usually used to correlate the vapor-liquid surface tension in the low and room temperature range (Ferguson and Kennedy, 1936; Guggenheim, 1945). Critical parameters T_c and V_c or intermolecular potential parameters σ and ε may be used to reduce the surface tension in accordance with the corresponding-states principle. This method was applied to nonpolar liquids applicable to the range of temperature away from the critical point (Brock and Bird, 1955; Curl and Pitzer, 1958). Miller (1963) suggested a method to relate the surface tension σ with the Riedel parameter α_c, while Riedel (1955) also proposed a similar correlation. To broaden this approach to polar fluids, Hakim et al. (1971) used the Stiel polar factor in their correlations. However, the general reliability was not known; the values of polar factors are available for only a few substances and estimated values of the surface tension σ are sensitive to the value of the chosen polar factor (Reid et al., 1987; Poling et al., 2001). Sivaraman et al. (1984) developed a correlation for predicting the surface tensions of organic compounds in the middle temperature range, which is, however, very sensitive to the values of the acentric factor, critical temperature, and critical pressure. Somayajulu (1988) proposed an equation with three substance-dependent parameters for the range of temperature from the triple point to the critical point. Perez-Lopez et al. (1992) used the gradient theory to predict the surface tension of nonpolar and polar fluids with a rather large deviation. There is still a need for an expression that could be applied over the entire range for all classes of substances. If the other corresponding-states parameters are introduced, for example, acentric factor and aspherical factor, the vapor-liquid surface tension may be correlated for a wide range of fluids including complex polar fluids (Sanchez, 1983; Reid et al., 1987; Escobedo and Mansoori, 1996; Miqueu et al., 2000, 2003; Xiang, 2001a, b, c, 2002, 2003). This chapter first reviews and introduces theories of the surface tension of liquids, then gives the derivation of the Macleod equation from statistical mechanics, finally describes the method and its application for the surface tension from the extended corresponding-states theory.

8.2 Critical-State Theory for the Surface Tension

8.2.1 Critical Behavior of the Surface Tension

The problem of critical phenomena in the surface tension of fluids may be defined as that of understanding the origin of the critical-point exponents μ and ν, and of determining their relation to each other and to the other indices that characterize a critical point. As a critical point is approached in a system in which two fluid phases are in equilibrium, the interface between the phases thickens and becomes diffuse, spreading ever further into the bulk phases until, at the critical point, it fills the whole of the volume, at which point the two phases have lost their separate identities and have merged into one homogeneous fluid phase. Associated with increasing the diffuseness of the interface is a gradual decrease in the interfacial tension, i.e., in the free energy of inhomogeneity per unit area of the interface, the latter phase then vanishing at the critical point. As the temperature T at which the two phases are in equilibrium approaches the critical temperature T_c, the interfacial tension σ is found experimentally to vanish proportionally to a power μ of $T_c - T$. The exponent μ is an important and characteristic critical-point index, with a value that is believed to be universal. Further, if κ is the reciprocal of the interface thickness, then as the critical point is approached, κ vanishes proportionally to a power of $T_c - T$. The exponent ν has a value that is believed to be universal $\nu = 0.63$.

The underlying ideas can be traced to Fisk and Widom (1969), who extended the theory of Van der Waals by incorporating the non-analytic behavior of thermodynamic properties near the critical point to predict the behavior of the surface tension near the critical point. Fisk and Widom concluded that universal ratios should exist between the amplitude σ_0, which characterizes the vanishing of the interfacial tension in the critical region, and the amplitude ξ_0, which characterizes the divergence of the correlation length, and a combination of amplitudes that characterize the non-analytic part of the free energy near the critical point. Although the ratios were predicted to be universal, the amplitudes themselves are not, and their values have no obvious connection with the critical parameters or the parameters for model intermolecular potentials. Stauffer *et al.* (1972) introduced the concept of two-scale-factor universality, which asserts that the singular part of the free energy belonging to a volume ξ^d, in which ξ is the correlation length and d the dimensionality, is a finite quantity, which is also universal for fluids. They concluded that separate universal ratios should exist between σ_0 and ξ_0 and between σ_0 and each of the amplitudes characterizing the divergence of the specific heat per unit volume. The existence of separate ratios leads to a very useful result: measurement of the interfacial tension, the correlation length, or the specific-heat divergence leads to knowledge of all three quantities on the coexistence curve and its extension.

8.2.2 Critical Universality of the Surface Tension

As noted by Moldover (1985), the customary power-law temperature dependence is used for the surface tension, the difference between the densities of coexisting phases, the isothermal compressibility, the specific heat, and the correlation length:

$$\sigma = \sigma_0 (-t)^\mu (1+...)$$

(8.1)

$$\Delta\rho = 2\rho_c B_0 (-t)^\beta [1 + B_1 (-t)^{\Delta_1} +...]$$

(8.2)

$$p_c^{-1}(\partial\rho/\partial\mu)_T^+ = \Gamma^+ \rho_c^2 t^{-\gamma}(1+...)$$

(8.3)

$$C_V^+ / R = A_0^+ |t|^{-\alpha} (1 + A_1^+ |t|^{\Delta_1} +...)$$

(8.4)

$$\xi^+ = \xi_0^+ |t|^{-\nu}(1+...)$$

(8.5)

where R is the gas constant, and T_c, ρ_c and p_c are the critical temperature, density, and pressure, respectively. The reduced temperature is defined by $t \equiv (T - T_c)/T_c$, except in the case of lower critical solution temperatures, where $t \equiv (T_c - T)/T_c$ is used. The superscripts + and – refer to the critical isochore or isopleth in the homogeneous and inhomogeneous regions, respectively. The specific heat C_V is the molar heat capacity at constant volume for liquid-vapor systems and the molar heat capacity at constant pressure and composition for liquid-liquid systems. μ in Eq. (8.3) is the chemical potential.

The capillary length is defined by

$$a^2 = 2\sigma /(\Delta\rho g),$$

(8.6)

where $g = 9.81 \text{m/s}^2$ is the acceleration due to gravity. The capillary length has the representation

$$a^2 = a_0^2 |t|^\phi (1+...).$$

(8.7)

The critical exponents are assumed to have the values calculated via the renormalization group (Albert, 1982; Le Guillou and Zinn-Justin, 1980; Moldover, 1985); namely,

$$\alpha = 0.110, \quad \nu = 0.630, \quad \gamma = 1.241, \quad \beta = 0.325.$$

(8.8)

In three dimensions, the scaling law for the interfacial tension gives the power in Eq. (8.1) as

$$\mu = 2\nu = 1.26.$$

(8.9)

From the definition of a^2 it follows that

$$\phi = \mu - \beta = 0.935 \,. \tag{8.10}$$

For the liquid-vapor systems, σ_0 was obtained from separate data for the capillary length and the coexisting densities. Thus, each amplitude ratio has two possible forms, one using σ_0 and one using a_0:

$$R^{\pm}_{\sigma A} \equiv \left[\frac{A^{\pm}_0 \rho_c N_0}{M} \right]^{2/3} \frac{k_B T_c}{\sigma_0} = \left[\frac{A^{\pm}_0 \rho_c N_0}{M} \right]^{2/3} \frac{k_B T_c}{a^2_0 B_0 \rho_c g} \,, \tag{8.11}$$

and

$$R^{\pm}_{\sigma\xi} \equiv \frac{\sigma_0 (\xi^+_0)^2}{k_B T_c} = \frac{a^2_0 B_0 \rho_c g (\xi^+_0)^2}{k_B T_c} \,, \tag{8.12}$$

where N_0 is Avagadro's constant, k_B is Boltzmann's constant, and M is the average molecular weight of the fluid system. Rowlinson and Widom (1982) presented an extensive treatment of interfaces near critical points where nonclassical critical exponents are incorporated, although its applicability is confined to the critical region (Levelt Sengers and Sengers, 1981).

8.3 Corresponding-States Surface Tension

8.3.1 Statistical-Mechanical Derivation of the Macleod Equation

As mentioned above, the Macleod equation expresses the surface tension of a liquid in equilibrium with its vapor as a function of the liquid- and vapor-phase densities as:

$$\sigma = K(\rho_1 - \rho_v)^4 \,, \tag{8.13}$$

where K is a substance-dependent parameter. Sugden (1924) modified Eq. (8.13) as follows:

$$\sigma = [P(\rho_1 - \rho_v)]^4 \,, \tag{8.14}$$

where $P = K^{1/4}$, which Sugden called the parachor, and described a method to estimate it based on molecular structure. The good performance and extreme simplicity of Eq. (8.14) have made it a very popular equation for the surface tension calculation, as reviewed by Escobedo and Mansoori (1996). It was shown that the parachor P is substance-dependent, is a weak function of temperature, and is almost temperature-independent in a wide temperature range (Macleod, 1923; Sugden, 1924; Quayle, 1953). Two methods have been used. The first starts with the classical thermodynamic expression, which relates the surface tension and the surface internal energy. The second (Boudh-Hir and Mansoori, 1990) begins with the statistical-mechanical definition of the surface tension. Using the first method (Fowler, 1937; Green, 1969), it has been shown that the surface tension is proportional to the fourth power of the difference in densities ($\rho_l - \rho_v$). Boudh-Hir and Mansoori (1990) derived the Macleod equation from statistical mechanics via the relation of the surface tension of a liquid in equilibrium with its vapor as a function of the one-particle densities.

Starting with the statistical-mechanical definition of the surface tension, Boudh-Hir and Mansoori (1990) have shown that, as a first approximation, the surface tension is given by the Macleod formula. As a result, the fourth power of the difference in densities ($\rho_l - \rho_v$) is obtained. However, they have shown that the constant K, which depends on the nature of the liquid under consideration, is not entirely independent of temperature. The only simplification they made was to consider the particles to interact via a generalized additive pairwise potential, which depends on the position and orientation of the molecules, i.e., the particles are not spherical and the potential is not necessary a radial function.

The statistical-mechanical expression for the surface tension derived by Boudh-Hir and Mansoori (1990) is

$$\sigma = [(kT/4)\tau^{4-2g}(z/z_c)\zeta(\tau, \rho_1, \rho_v)](\rho_1 - \rho_v)^4 \qquad (8.15)$$

where

$$\zeta(\tau, \rho_1, \rho_v) = \int \partial_{z_1}\chi(1;\xi)e^{\rho_c\tau\chi(1;\xi)}\partial_{z_2}\chi(2;\xi)e^{\rho_c\tau\chi(2;\xi)}c(1,2)(r_{12}^2 - z_{12}^2)dz_1 d\Omega_1 dr_2 d\Omega_2 \qquad (8.16)$$

Here k is the Boltzmann constant; T is the temperature; $\tau = (1 - T/T_c)$; g is an exponent; $z = (2\pi mkT/h^2)^{1/2}e^{\mu/kT}$ is the activity; the subscript c denotes the value of the activity at the critical temperature (i.e., $z = (2\pi mkT_c/h^2)^{1/2}e^{\mu/kT_c}$); μ is the chemical potential; h is the Planck constant; $\rho_c\tau\chi(i;\xi) = \Delta c(i)$; ρ_c is the critical density; $\Delta c(i) = [c(i) - c_c(i)]$ is the difference of the value of the one-particle direct correlation function at the temperature from that at the critical temperature. $\chi(i;\xi)$ is given by the following expression:

$$\chi(i;\xi) = \int c(i,j;\xi)[-g_c v(i,j)e^{-g_c w(i,j)} + g_c\mu - 3/2]e^{\Delta c(j;\xi)} , \qquad (8.17)$$

where $c(i,j;\xi)$ is the two-particle direct correlation function; $w(i,j)$ is a pairwise potential and $v(i,j)$ the mean force potential; ξ is an order parameter which depends on τ. $\xi = 0$ corresponds to the system at the critical temperature, while $\xi = 1$ is associated with the system at the temperature of interest. g_c is the value of the exponent g at the critical temperature. Boudh-Hir and Mansoori (1990) derived Eqs. (8.15) to (8.17) in detail. As indicated by Escobedo and Mansoori (1996), Eq.(8.15) is general and may be applied to any fluid if the additive pairwise potential is satisfied. Although the form of the first term of the surface tension expansion remains the same near the critical temperature. The parameter K in Macleod's formula is a weak function of the temperature and becomes very sensitive to this parameter when it increases and tends to its critical temperature T_c. Consequently, Eq. (8.15) is a suitable expression to represent the surface tension.

8.3.2 Corresponding States of the Modified Macleod Equation

The Macleod equation has been proven to work very well for many substances over a wide range of temperatures. There have been a number of efforts made to justify the success of Macleod's formula from a theoretical basis (Fowler, 1937; Green, 1969; Henderson, 1980; Rowlinson and Widom, 1982; Boudh-Hir and Mansoori, 1990). Nonetheless, deviations with respect to temperature have generally been observed. Thus efforts have been made to derive the parachor as a function of temperature (Escobedo and Mansoori, 1996). The surface tension of various substances can be represented by a method that is based on the extended corresponding-states theory and a reference equation that is valid over the entire range. According to the temperature dependence of surface tension $\sigma \propto |t|^{1.26}$ and of the difference between the densities of coexisting phases given by Eq. (8.2), the following equation is proposed for the surface tension, which is modified from Xiang (2003):

$$\sigma / \sigma_0 = \left[s_0 + s_1 (1/T_r - 1)^{1/3} + s_2 (1/T_r - 1) \right](1/T_r - 1)^{-0.04} (\rho_{l,r} - \rho_{v,r})^4, \quad (8.18)$$

where $\sigma_0 = kT_c (N_A \rho_c / M)^{2/3}$. Eq. (8.18) may describe the surface tension of simple, nonpolar, polar, hydrogen-bonding, and associating substances. The substance-dependent parameters s_0, s_1, and s_2 in Eq. (8.18) are generalized by means of the extended corresponding-states theory:

$$s_0 = s_{00} + s_{01}\omega + s_{02}\theta,$$
$$s_1 = s_{10} + s_{11}\omega + s_{12}\theta,$$
$$s_2 = s_{20} + s_{21}\omega + s_{22}\theta. \tag{8.19}$$

The coefficients s_{ij} of Eq. (8.19) given in Table 8.1 are obtained by fitting the experimental data of argon, the weakly nonspherical molecules, such as ethane, propane, 1,1,1,2-tetrafluoroethane and 1,1-difluoroethane, and the highly nonspherical molecule, such as water. The critical parameters and acentric factors of these substances and the other substances listed for comparison are given in Table 8.2, with the data sources listed in Table 8.3. These experimental values are the most accurate up to date and in general have been widely used to test proposed models.

Table 8.1 General Coefficients of Eq. (8.39)

s_{00}	0.03751642	s_{10}	-0.00983666	s_{20}	0.004674589
s_{01}	-0.01990581	s_{11}	-0.02497753	s_{21}	-0.01705001
s_{02}	-2.685506	s_{12}	-1.438909	s_{22}	1.597544

8.3.3 Comparison with Experimental Data and Existing Methods

As shown in Figs. 8.1 and 8.2, the extended corresponding-states theory, Eqs. (8.18) and (8.19), represent surface-tension data for some representative real fluids over their entire temperature ranges. In the figures, only one symbol is used for all of the data sets for each molecule for the sake of clarity. For most cases, the agreement between the present theory and the data is generally comparable to the agreement between the different data sets for each substance.

It can be seen that the extended corresponding-states theory describes the surface tension to within the uncertainties of data over their entire temperature ranges. In the high-temperature range, this property is subject to large uncertainties, where the value approaches zero at the critical point, as a result, the uncertainty of experimental data is large. The experimental difficulties in obtaining accurate high-temperature data mean that experimental results are scarce and have great uncertainty, while the extended corresponding-states theory accurately represents this physical behavior and also predicts it well at low temperatures.

While the three-parameter corresponding-states methods of Curl and Pitzer (1958) and consequently of Zuo and Stenby (1997), based on the work of Rice and Teja (1982), are satisfactory for nonpolar liquids, they are not satisfactory for compounds that are strong hydrogen-bonding substances, such as alcohols, acids, and water. This is illustrated for water and methanol in Fig. 8.3 from the corresponding-states theory of Pitzer et al. In contrast, the extended corresponding-states theory is capable of representing the surface tension of highly polar, hydro-bonding and associating substances, such as difluoromethane, 1, 1, 1, 2-tetrafluoroethane, 1, 1-difluoroethane, water, and methanol as shown in Fig. 8.2.

As reviewed by Poling et al. (2001), other approaches to surface tension prediction include a method by Escobedo and Mansoori (1996) in which the parachor is related to the refractive index, and a method by Hugill and Van Welsenes (1986), which employs a corresponding-states expression for the parachor. Both of these methods work well for nonpolar fluids but can lead to large errors for polar compounds.

For the difference between the densities of coexisting phases, B_0 in Eq. (8.2) is correlated by the extended corresponding-states theory as $B_0 = 1.47994 + 0.6652899\omega + 68.576810$. It can be seen that for highly polar substances, the present theory agrees well with recent highly accurate data over the entire region. When the surface tension approaches the critical point as a function of temperature, a good representation of the surface tension is shown in Fig. 8.4.

Table 8.2 Molar Mass M, Critical Temperature T_c, Critical pressure p_c, Critical density ρ_c, Acentric Factor ω, and Aspherical Factor θ

Substance	M $(\mathrm{kg.kmol^{-1}})$	T_c (K)	p_c (kPa)	ρ_c $(\mathrm{kg.m^{-3}})$	ω	$\theta \times 10^3$
Argon	39.948	150.69	4863	535	0.000	0.000
Nitrogen	28.013	126.19	3395	313	0.037	0.000
Methane	16.043	190.56	4598	162	0.011	0.007
Methyl chloride	50.488	416.30	6700	363	0.153	0.432
Carbonyl sulfide	60.070	378.8	6350	440	0.105	0.217
Carbon dioxide	44.010	304.13	7377	468	0.225	0.245
Acetylene	26.038	308.3	6140	231	0.190	0.400
Ethylene	28.054	282.35	5041	215	0.086	0.096
Ethane	30.070	305.32	4872	206	0.099	0.097
Propane	44.097	369.83	4248	220	0.152	0.171
Benzene	78.114	562.05	4895	305	0.210	0.472
Pentane	72.150	469.7	3370	232	0.251	0.468
Heptane	100.204	540.2	2740	234	0.350	0.827
Octane	114.231	568.7	2490	234	0.398	1.087
Decane	142.86	617.7	2110	228	0.491	1.061
CH_2F_2	52.020	351.26	5780	430	0.277	2.560
CCl_2CHF_3	152.93	456.83	3662	550	0.282	0.481
CF_3CHF_2	120.02	339.17	3620	570	0.306	0.388
CF_3CH_2F	102.03	374.18	4055	511	0.327	0.885
CH_4CF_2	66.050	386.41	4516	369	0.275	1.474
Dimethyl ether	46.069	400.0	5330	270	0.197	0.274
Diethyl ether	74.123	466.6	3640	258	0.281	0.420

Ethyl acetate	88.106	523.2	3880	300	0.367	0.787
Nitrous oxide	44.013	309.6	7240	452	0.165	0.260
Sulfur dioxide	64.063	430.8	7880	524	0.256	0.443
Methyl acetylene	40.065	402.4	5630	244	0.215	0.188
Ammonia	17.031	405.4	11345	234	0.256	2.028
Water	18.015	647.1	22050	325	0.344	3.947
Methanol	32.042	512.0	8000	270	0.560	4.486
Ethanol	46.069	514.0	6137	275	0.644	2.444
Propanol	60.096	536.8	5170	274	0.623	1.292
2-Butanol	74.123	563.1	4420	269	0.593	0.892
Decanol	158.285	687	2220	263	0.629	3.146
Acetone	58.080	508.1	4700	270	0.308	2.569
Acetic acid	60.052	593.0	5786	330	0.450	5.844

Table 8.3 Data Source of the Surface Tension for Some Typical Substances

Substance	Ref(s).
Argon	Baidakov and Skripov, 1982; Stansfield, 1958
Nitrogen	Stansfield, 1958
Methane	Fuks and Bellemans, 1966; Soares et al., 1986
Ethane	Baidakov and Sulla, 1987; Katz and Saltman, 1939; Leadbetter et al., 1964; Maass and Wright, 1921; Soares et al., 1986
Propane	Katz and Saltman, 1939
n-Pentane	Grigoryev et al., 1992
n-Hexane	Grigoryev et al., 1992
n-Heptane	Grigoryev et al., 1992
n-Octane	Grigoryev et al., 1992
Sulfur hexafluoride	Rathjen and Straub, 1977
Difluoromethane	Froeba et al., 2000; Okada and Higashi, 1995; Schmidt and Moldover, 1994; Zhu and Lu, 1994
1,1-Dichloro-2,2,2-trifluoroethane	Chae et al., 1990
1,1,1,2-Tetrafluoroethane	Chae et al., 1990; Froeba et al., 2000
1,1-Difluoroethane	Chae et al., 1990a; Froeba et al., 2000; Okada and Higashi, 1995
Water	IAPWS, 1994
Methanol	Strey and Schmeling, 1983

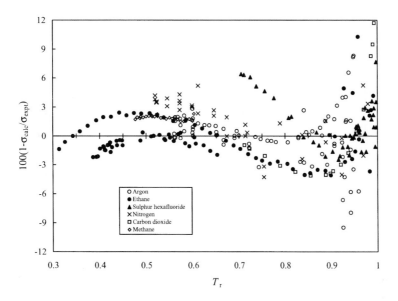

Fig. 8.1 Comparison of surface-tension data for nonpolar molecules with the values calculated from the extended corresponding-states theory, Eqs. (9.18) and (9.19).

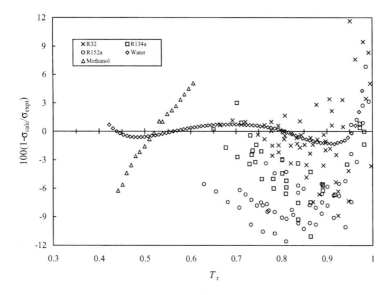

Fig. 8.2 Comparison of surface-tension data for highly polar molecules with the values calculated from the extended corresponding-states theory, Eqs. (9.18) and (9.19).

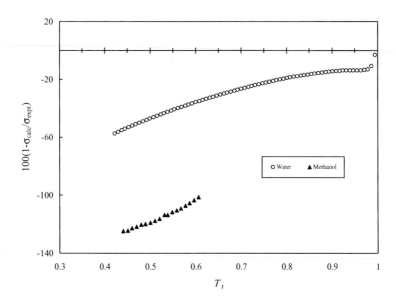

Fig. 8.3 Comparison of surface-tension data for highly polar molecules with the values calculated from the corresponding-states theory of Pitzer et al. (Curl and Pitzer, 1958)

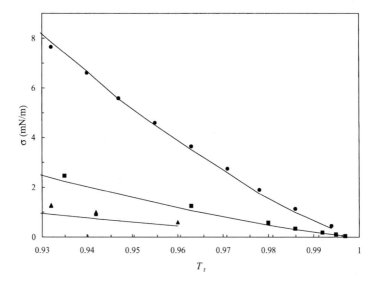

Fig. 8.4 Comparison of the surface tension as a function of temperature $\tau^{1.26}$ in the vicinity of the critical point derived from Eq. (8.18) and (9.19) with the values calculated from the extended corresponding-states theory. (▲) argon; (■) difluoromethane; (●)water

References

Albert, D. Z., 1982, *Phys. Rev. B* **25**: 4810.

Baidakov, V. G. and V. P. Skripov, 1982, *Kolloidn. Zh.* **44**: 409.

Baidakov, V. G. and I. I. Sulla, 1987, *Ukr. Fiz. Zh.* **32**: 885.

Boudh-Hir, M.-E. and G. A. Mansoori, 1990, *J. Phys. Chem.* **94**: 8362.

Brock, J. R. and R. B. Bird, 1955, *AIChE J.* **1**: 174.

Chae, H. B., J. W. Schmidt, and M. R. Moldover, 1990, *J. Chem. Eng. Data* **35**: 6.

Chae, H. B., J. W. Schmidt, and M. R. Moldover, 1990a, *J. Phys. Chem.* **94**: 8840.

Curl, R. F. and K. S. Pitzer, 1958, *Ind. Eng. Chem.* **50**: 265.

Edison, T. A. and J. V. Sengers, 1999, *Int. J. Refrig.* **22**: 365.

Escobedo, J. and G. A. Mansoori, 1996, *AIChE J.* **42**: 1425.

Ferguson, A. and S. J. Kennedy, 1936, *Trans. Faraday Soc.* **32**: 1474.

Fisk, S. and B. Widom, 1969, *J. Chem. Phys.* **50**: 3219.

Fowler, R. H., 1937, *Proc. Roy. Soc.* **159A**: 229.

Froeba, A. P., S. Will, and A. Leipertz, 2000, *Int. J. Thermophys.* **21**: 1225.

Fuks, S. and A. Bellemans, 1966, *Physica* **32**: 594.

Green, H. S., 1969, The Molecular Theory of Liquids (Dover, New York).

Grigoryev, B. A., B. V. Nemzer, D. S. Kurumov, and J. V. Sengers, 1992, *Int. J. Thermophys.* **13**: 453.

Guggenheim, E. A., 1945, *J. Chem. Phys.* **13**: 253.

Hakim, D. I., D. Steinberg, and L. I. Stiel, 1971, *Ind. Eng. Chem. Fundam.* **10**: 174.

Henderson, J. R., 1980, *Mol. Phys.* **39**: 709.

Hugill, J. A. and A. J. van Welsenes, 1986, *Fluid Phase Equilib.* **29**: 383.

IAPWS, 1994, Release on Surface Tension of Ordinary Water Substance, International Association for the Properties of Water and Steam.

Katz, D. L. and W. Saltman, 1939, *Ind. Eng. Chem.* **31**: 91.

Laine, P., 1953, *Kaeltetech.* **6**: 173.

Leadbetter, A. J., D. J. Taylor, and B. Vincent, 1964, *Can. J. Chem.* **42**: 2930.

Le Guillou, J. C. and J. Zinn-Justin, 1980, *Phys. Rev. B* **21**: 3976.

Levelt-Sengers, J. M. H. and J. V. Sengers, 1981, Perspectives in Statistical Physics (North-Holland, Amsterdam).

Maass, M. and C. H. Wright, 1921, *J. Am. Chem. Soc.* **43**: 1098.

Macleod, D. B., 1923, *Trans. Faraday Soc.* **19**: 38.

Miller, D. G., 1963, *Ind. Eng. Chem. Fundam.* **2**: 78.

Miqueu, C., D. Broseta, J. Satherley, B. Mendiboure, J. Lachaise, and A. Graciaa, 2000, *Fluid Phase Equil.* **172**: 169.

Miqueu, C., B. Mendiboure, A. Graciaa, and J. Lachaise, 2003, *Fluid Phase Equil.* **207**: 225; **212**: 363.

Moldover, M. R., 1985, *Phys. Rev. A* **31**: 1022.

Okada, M. and Y. Higashi, 1995, *Int. J. Thermophys.* **16**: 791.

Perez-Lopez, J. H., L. J. Gonzales-Ortiz, M. A. Leiva, and J. E. Puig, 1992, *AIChE J.* **38**: 753.

Poling, B. E., J. M. Prausnitz, and J. P. O'Connell, 2001, The Properties of Gases and Liquids, 5th ed. (McGraw-Hill, New York).

Quayle, O. R., 1953, *Chem. Rev.* **53**: 439.

Rathjen, W. and J. Straub, 1977, in *Proc. Seventh Symp. Thermophys. Prop.* A. Cezairliyan ed. (ASME, New York), pp 839-850.

Reid, R. C., J. M. Prausnitz, and B. E. Poling, 1987, The Properties of Gases and Liquids, 4th ed. (McGraw-Hill, New York).

Rice, P. and A. S. Teja, 1982, *J. Colloid. Interface Sci.* **86**: 158.

Riedel, L., 1954, *Chem. Ing. Tech.* **26**: 83.

Riedel, L., 1955, *Chem. Ing. Tech.* **27**: 209.

Rowlinson, J. S. and B. Widom, 1982, Molecular Theory of Capillarity (Clarendon Press, Oxford).

Sanchez, I. C. 1983, *J. Chem. Phys.* **79**: 405.

Schmidt, J. W. and M. R. Moldover, 1994, *J. Chem. Eng. Data* **39**: 39.

Sivaraman, A., J. Zega, and R. Kobayashi, 1984, *Fluid Phase Equil.* **18**: 225.

Soares, V. A. M., B. de J. V. S. Almeida, I. A. McLure, and R. A. Higgins, 1986, *Fluid Phase Equilib.* **32**: 9.

Somayajulu, G. R., 1988, *Int. J. Thermophys.* **9**: 559.

Stansfield, D., 1958, *Proc. Phys. Soc. London* **72**: 854.

Stauffer, D., M. Ferer, and M. Wortis, 1972, *Phys. Rev. Lett.* **29**: 345.

Strey, R. and T. Schmeling, 1983, *Ber. Bunsenges. Phys. Chem.* **87**: 324.

Sugden, S., 1924, *J. Chem. Soc.* **125**: 32, 1177.

Tang, S., J. V. Sengers, and Z. Y. Chen, 1991, *Physica A* **179**: 344.

Van der Waals, J. D., 1873, On the Continuity of the Gaseous and Liquid States, Rowlinson, J. S., ed. Studies in Statistical Mechanics, Vol.14 (North-Holland, Amsterdam, 1988).

Van der Waals, J. D., 1894, *Z. Phys. Chem.* **13**: 657, 716.

Wyczalkowska, A. K., Kh. S. Abdulkadirova, M. A. Anisimov, and J. V. Sengers, 2000, *J. Chem. Phys.* **113**: 4985.

Xiang, H. W., 2001a, *Int. J. Thermophys.* **22**: 919.

Xiang, H. W., 2001b, *J. Phys. Chem. Ref. Data* **30**: 1161.

Xiang, H. W., 2001c, *Fluid Phase Equil.* **187**: 221.

Xiang, H. W., 2002, *Chem. Eng. Sci.* **57**: 1439.

Xiang, H. W., 2003, Thermophysicochemical Properties of Fluids: Corresponding-States Principle and Practise (Science Press, Beijing), in Chinese.

Zhu, M.-S. and C.-X. Lu, 1994, *J. Chem. Eng. Data* **39**: 205.

Zuo, Y.-X. and E. H. Stenby, 1997, *Can. J. Chem. Eng.* **75**: 1130.

Index